The Number Sense

The Number Sense

How the Mind Creates Mathematics

Stanislas Dehaene

New York Oxford
Oxford University Press
1997

OXFORD UNIVERSITY PRESS

Oxford New York
Athens Auckland Bangkok Bogotá Bombay
Buenos Aires Calcutta Cape Town Dar es Salaam
Delhi Florence Hong Kong Istanbul Karachi
Kuala Lumpur Madras Madrid Melbourne
Mexico City Nairobi Paris Singapore
Taipei Tokyo Toronto Warsaw

and associated companies in
Berlin Ibadan

Published by Oxford University Press, Inc.,
198 Madison Avenue, New York, New York 10016-4314

Oxford is a registered trademark of Oxford University Press

Library of Congress Cataloging-in-Publication Data

Dehaene, Stanislas.
The number sense : how the mind creates mathematics
/ by Stanislas Dehaene.
p. cm. Includes bibliographical references and index.
ISBN 0-19-511004-8
1. Number concept. 2. Mathematics—Study and teaching—
Psychological aspects. 3. Mathematical ability. I. Title.
QA141.D44 1997
510'.1'9—dc20 96-53840

1 3 5 7 9 8 6 4 2

Printed in the United States of America
on acid-free paper

Contents

Preface

We are surrounded by numbers. Etched on credit cards or engraved on coins, printed on pay checks or aligned on computerized spread sheets, numbers rule our lives. Indeed, they lie at the heart of our technology. Without numbers, we could not send rockets roaming the solar system, nor could we build bridges, exchange goods, or pay our bills. In some sense, then, numbers are cultural inventions only comparable in importance to agriculture or to the wheel. But they might have even deeper roots. Thousands of years before Christ, Babylonian scientists used clever numerical notations to compute astronomical tables of amazing accuracy. Tens of thousands of years prior to them, Neolithic men recorded the first written numerals by engraving bones or by painting dots on cave walls. And as I shall try to convince you later on, millions of years earlier still, long before the dawn of humankind, animals of all species were already registering numbers and entering them into simple mental computations. Might numbers, then be almost as old as life itself? Might they be engraved in the very architecture of our brains? Do we all possess a "number sense," a special intuition that helps us make sense of numbers and mathematics?

Fifteen years ago, as I was training to become a mathematician, I became fascinated by the abstract objects I was taught to manipulate, and above all, by the simplest of them—numbers. Where did they come from? How was it possible for my brain to understand them? Why did it seem so difficult for most people to master them? Historians of science and philosophers of mathematics had provided some tentative answers, but to a scientifically oriented mind their speculative and contingent character was unsatisfactory. Furthermore, scores of intriguing facts about numbers and mathematics were left unanswered in the books I knew of. Why did all languages have at least some number names? Why did everybody seem to find multiplications by seven, eight, or nine particularly hard to learn? Why couldn't I seem to recognize more than four objects at a glance? Why were there ten boys for one girl in the high-level mathematics classes I was attending? What tricks allowed lightning calculators to multiply two three-digit numbers in a few seconds?

As I learned increasingly more about psychology, neurophysiology, and computer science, it became obvious that the answers had to be looked for, not in history books, but in the very structure of our brains—the organ that enables us to create mathematics. It was an exciting time for a mathematician to turn to cognitive neuroscience. New experimental techniques and amazing results seemed to appear every month. Some revealed that animals could do simple arithmetic. Others asked whether babies had any notion of 1 plus 1. Functional imaging tools were also becoming available that could visualize the active circuits of the human brain as it calculates and solves arithmetical problems. Suddenly, the psychological and cerebral bases of our number sense were open to experimentation. A new field of science was emerging: mathematical cognition, or the scientific inquiry into how the human brain gives rise to mathematics. I was lucky enough to become an active participant in this quest. This book provides a first glance at this new field of research that my colleagues in Paris and several research teams throughout the world are still busy developing.

I am indebted to many people for helping me complete the transition from mathematics to neuropsychology. First and foremost, my research program on arithmetic and the brain could never have developed without the generous assistance of three outstanding teachers, colleagues, and friends who deserve very special thanks: Jean-Pierre Changeux in neurobiology, Laurent Cohen in neuropsychology, and Jacques Mehler in cognitive psychology. Their support, advice, and often direct contribution to the work described here have been of invaluable help.

Through the years, I have also benefited from the advice of many other eminent scientists. Mike Posner, Don Tucker, Michael Murias, Denis Le Bihan, André Syrota, and Bernard Mazoyer shared with me their in-depth knowledge of brain imaging. Emmanuel Dupoux, Anne Christophe, and Christophe Pallier advised me in psycholinguistics. I am also grateful for ground-shaking debates with Rochel Gelman and Randy Gallistel and for judicious remarks by Karen Wynn, Sue Carey, and Josiane Bertoncini on child development.

The late professor Jean-Louis Signoret had introduced me to the fascinating domain of neuropsychology. Subsequently, numerous discussions with Alfonso Caramazza, Michael McCloskey, Brian Butterworth, and Xavier Seron greatly enhanced my understanding of this discipline. Xavier Jeannin and Michel Dutat, finally, assisted me in programming my experiments.

I also gratefully acknowledge the crucial contribution of several students to my research: Rokny Akhavein, Serge Bossini, Florence Chochon, Pascal Giraux, Markus Kiefer, Etienne Koechlin, Lionel Naccache, and Gérard Rozsavolgyi.

The preparation of this book greatly benefited from the close scrutiny of Brian Butterworth, Robbie Case, Markus Giaquinto, and Susana Franck for the English edition, and of Jean-Pierre Changeux, Laurent Cohen, and Ghislaine Dehaene-Lambertz for the French edition. Warm thanks go also to Joan Bossert, my editor at Oxford University Press, John Brockman, my agent, and Odile Jacob, my French editor. Their trust and support was very precious.

I would also like to thank the publishers and authors who kindly granted me the permission to reproduce the figures and quotes used in this book. Special thanks go to Gianfranco Denes for drawing my attention to the remarkable section of Ionesco's *Lesson* that is cited in Chapter 8.

Last but not least, a word of thanks cannot suffice to express my feelings for my family, Ghislaine, Oliver, David, and Guillaume, who patiently supported me during the long months spent exploring and writing about the universe of numbers. This book is dedicated to them.

S. D.
Piriac, France
August 1996

The Number Sense

Introduction

> Any poet, even the most allergic to mathematics, has to count up to twelve in order to compose an alexandrine.
>
> *Raymond Queneau*

As I first sat down to write this book, I was faced with a ridiculous problem of arithmetic: If this book is to have 250 pages and nine main chapters, how many pages will each chapter have? After thinking hard, I came to the conclusion that each should have slightly fewer than 30 pages. This took me about five seconds, not bad for a human, yet an eternity compared to the speed of any electronic calculator. Not only did my calculator respond instantaneously, but the result it gave was accurate to the tenth decimal: 27.7777777778!

Why is our capacity for mental calculation so inferior to that of computers? And how do we reach excellent approximations such as "slightly fewer than 30" without resorting to an exact calculation, something that is beyond the best of electronic calculators? The resolution of these nagging questions, which is the subject matter of this book, will confront us with even more challenging riddles:

- Why is it that after so many years of training, the majority of us still do not know for sure whether 7 times 8 is 54 or 64 . . . or is it 56?
- Why is our mathematical knowledge so vulnerable that a small cerebral lesion is enough to abolish our sense of numbers?

- How can a five-month-old baby know that 1 plus 1 equals 2?
- How is it possible for animals without language, such as chimpanzees, rats, and pigeons, to have some knowledge of elementary arithmetic?

My hypothesis is that the answers to all these questions must be sought at a single source: the structure of our brain. Every single thought we entertain, every calculation we perform, results from the activation of specialized neuronal circuits implanted in our cerebral cortex. Our abstract mathematical constructions originate in the coherent activity of our cerebral circuits and of the millions of other brains preceding us that helped shape and select our current mathematical tools. Can we begin to understand the constraints that our neural architecture impose on our mathematical activities?

Evolution, ever since Darwin, has remained the reference for biologists. In the case of mathematics, both biological and cultural evolution matter. Mathematics is not a static and God-given ideal, but an ever-changing field of human research. Even our digital notation of numbers, as obvious as it may seem now, is the fruit of a slow process of invention over thousands of years. The same holds for the current multiplication algorithm, the concept of square root, the sets of real, imaginary, or complex numbers, and so on. All still bear scars of their difficult and recent birth.

The slow cultural evolution of mathematical objects is a product of a very special biological organ, the brain, that itself represents the outcome of an even slower biological evolution governed by the principles of natural selection. The same selective pressures that have shaped the delicate mechanisms of the eye, the profile of the hummingbird's wing, or the minuscule robotics of the ant have also shaped the human brain. From year to year, species after species, ever more specialized mental organs have blossomed within the brain to better process the enormous flux of sensory information received, and to adapt the organism's reactions to a competitive or even hostile environment.

One of the brain's specialized mental organs is a primitive number processor that prefigures, without quite matching it, the arithmetic that is taught in our schools. Improbable as it may seem, numerous animal species that we consider stupid or vicious, such as rats and pigeons, are actually quite gifted at calculation. They can represent quantities mentally and transform them according to some of the rules of arithmetic. The scientists who have studied these abilities believe that animals possess a mental module, traditionally called the "accumulator," that can hold a register of various quantities. We shall see later how rats exploit this mental accumulator to distinguish series of two, three, or four sounds or to compute approximate additions of two quantities. The accumulator mechanism opens up a new dimension of sensory perception through which the cardinal of a set of objects can be perceived just as easily as their color, shape, or position. This "number sense" provides animals and humans alike with a direct intuition of what numbers mean.

Tobias Dantzig, in his book exalting "number, the language of science," underlined the primacy of this elementary form of numerical intuition: "Man, even in the lower stages of development, possesses a faculty which, for want of a better name, I shall call *Number Sense*. This faculty permits him to recognize that something has changed in a small collection when, without his direct knowledge, an object has been removed or added to the collection."

Dantzig wrote these words in 1954, when psychology was dominated by Jean Piaget's theory, which denied young children any numerical abilities. It took twenty more years before Piagetian constructivism was definitely refuted and Dantzig's insight was confirmed. All people possess, even within their first year of life, a well-developed intuition about numbers. Later we consider in some detail the ingenious experiments which demonstrate that human babies, far from being helpless, already know right from birth some fragments of arithmetic comparable to the animal knowledge of number. Elementary additions and subtractions are already available to six-month-old babies!

Let there be no misunderstanding. Obviously, only the adult *Homo sapiens* brain has the power to recognize that 37 is a prime number or to calculate approximations of the number π. Indeed, such feats remain the privilege of only a few humans in a few cultures. The baby brain and *a fortiori* the animal brain, far from exhibiting our mathematical flexibility, work their minor arithmetical miracles only within quite limited contexts. In particular, their accumulator cannot handle discrete quantities, but only continuous estimates. Pigeons will never be able to distinguish 49 from 50, because they cannot represent these quantities other than in an approximate and variable fashion. For an animal, 5 plus 5 does not make 10, but only *about 10*: maybe 9, 10, or 11. Such poor numerical acuity, such fuzziness in the internal vision of numbers, prevents the emergence of exact arithmetical knowledge in animals. By the very structure of their brains, they are condemned to an approximate arithmetic.

Humans, however, have been endowed by evolution with a supplementary competence: the ability to create complex symbol systems, including spoken and written language. Words or symbols, because they can separate concepts with arbitrarily close meanings, allow us to move beyond the limits of approximation. Language allows us to label infinitely many different numbers. These labels, the most evolved of which are the Arabic numerals, can symbolize and discretize any continuous quantity. Thanks to them, numbers that may be close in quantity but whose arithmetical properties are very different can be distinguished. Only then can the invention of purely formal rules for comparing, adding or dividing two numbers be conceived. Indeed, numbers acquire a life of their own, devoid of any direct reference to concrete sets of objects. The scaffolding of mathematics can then rise, ever higher, ever more abstract.

This raises a paradox, however. Our brains have remained essentially unchanged since *Homo sapiens* first appeared 100,000 years ago. Our genes, indeed, are condemned to a slow and minute evolution, dependent on the occurrence of chance mutations. It takes thousands of aborted attempts before a favorable mutation, one worthy of being passed on to coming generations, emerges from the noise. In contrast, cultures evolve through a much faster process. Ideas, inventions, progress of all kinds can spread to an entire population through language and education as soon as they have germinated in some fertile mind. This is how mathematics, as we know it today, has emerged in only a few thousand years. The concept of number, hinted at by the Babylonians, refined by the Greeks, purified by the Indians and the Arabs, axiomatized by Dedekind and Peano, generalized by Galois, has never ceased to evolve from culture to culture—obviously, without requiring any modification of the mathematician's genetic material! In a first approximation, Einstein's brain is no different from that of the master who, in the Magdalenian, painted the Lascaux cave. At elementary school, our children learn modern mathematics with a brain initially designed for survival in the African savanna.

How can we reconcile such biological inertia with the lightning speed of cultural evolution? Thanks to extraordinary modern tools such as positron emission tomography or functional magnetic resonance imaging, the cerebral circuits that underlie language, problem solving, and mental calculation can now be imaged in the living human brain. We will see that when our brain is confronted with a task for which it was not prepared by evolution, such as multiplying two digits, it recruits a vast network of cerebral areas whose initial functions are quite different, but which may, together, reach the desired goal. Aside from the approximate accumulator that we share with rats and pigeons, our brain probably does not contain any "arithmetical unit" predestined for numbers and math. It compensates this shortcoming, however, by tinkering with alternative circuits that may be slow and indirect, but are more or less functional for the task at hand.

Cultural objects—for instance written words or numbers—may thus be considered as parasites that invade cerebral systems initially destined to a quite different use. Occasionally, as in the case of word reading, the parasite can be so intrusive as to completely replace the previous function of a given brain area with its own. Thus, some brain areas that, in other primates, seem to be dedicated to the recognition of visual objects acquire in the literate human a specialized and irreplaceable role in the identification of letter and digit strings.

One cannot but marvel at the flexibility of a brain that can, depending on context and epoch, plan a mammoth hunt or conceive of a demonstration of Fermat's last theorem. However, this flexibility should not be overestimated. Indeed, my contention is that it is precisely the assets and the limits of our cerebral circuits that determine the strong and weak points of our mathematical abilities. Our brain, like that of the rat, has been endowed since time immemorial with an intuitive representation of quantities. This is why we are so gifted for approximation, and why it seems so obvious to us that 10 is larger than 5. Conversely, our memory, unlike that of the computer, is not digital but works by association of ideas. This is probably the reason why we have such a hard time remembering the small number of equations that make up the multiplication table.

Just as the budding mathematician's brain thus lends itself more or less easily to the requirements of mathematics, mathematical objects also evolve to match our cerebral constraints increasingly well. The history of mathematics provides ample evidence that our concepts of number, far from being frozen, are in constant evolution. Mathematicians have worked hard for centuries to improve the usefulness of numerical notations by increasing their generality, their fields of application, and their formal simplicity. In doing so, they have unwittingly invented ways of making them fit the constraints of our cerebral organization. Though a few years of education now suffice for a child to learn digital notation, we should not forget that it took centuries to perfect this system before it became child's play. Some mathematical objects now seem very intuitive only because their structure is well adapted to our brain architecture. On the other hand, a great many children find fractions very difficult to learn because their cortical machinery resists such a counterintuitive concept.

If the basic architecture of our brain imposes such strong limits on our understanding of arithmetic, why do a few children thrive on mathematics? How have outstanding mathematicians such as Gauss, Einstein, or Ramanujan attained such extraordinary familiarity with mathematical objects? And how do some idiot savants with an IQ of 50 manage to become experts in mental calculation? Do we have to suppose that some people started in life with a particular brain architecture or a biological predisposition to become geniuses? A careful examination of this supposition will show us that this is unlikely. At present, at any rate, very little evidence exists that great mathematicians and calculating prodigies have been endowed with an exceptional neurobiological structure. Like the rest of us, experts in arithmetic have to struggle with long calculations and abstruse mathematical concepts. If they succeed, it is only because they devote a considerable time to this topic and eventually invent well-tuned algorithms and clever shortcuts that any of us could learn if we tried and that are carefully devised to take advantage of our brain's assets and get round its limits. What is special about them is their disproportionate and relentless passion for numbers and mathematics, occasionally fueled by their inability to entertain normal relations with other fellow humans, a cerebral disease called autism. I am convinced that children of equal initial abilities may become excellent or hopeless at mathematics depending on their love or hatred of the subject. Passion breeds talent—and parents and teachers therefore have a considerable responsibility in developing their children's positive or negative attitudes toward mathematics.

In *Gulliver's Travels*, Jonathan Swift describes the bizarre teaching methods used at the mathematics school of Lagado, in Balnibarbi Island:

> I was at the mathematical school, where the master taught his pupils after a method scarcely imaginable to us in Europe. The proposition and demonstration were fairly written on a thin wafer, with ink composed of a cephalic tincture. This the student was to swallow upon a fasting stomach, and for three days following eat nothing but bread and water. As the wafer digested, the tincture mounted to his brain, bearing the proposition along with it. But the success hath not hitherto been answerable, partly by some error in the *quantum* or composition, and partly by the perverseness of lads, to whom this bolus is so nauseous, that they generally steal aside, and discharge it upwards before it can operate; neither have they been yet persuaded to use so long an abstinence as the prescription requires.

Although Swift's description reaches the height of absurdity, his basic metaphor of learning mathematics as a process of assimilation has an undeniable truth. In the final analysis, all mathematical knowledge is incorporated into the biological tissues of the brain. Every single mathematics course that our children take is made possible by the modifications of millions of their synapses, implying widespread gene expression and the formation of billions of molecules of neurotransmitters and receptors, with modulation by chemical signals reflecting the child's level of attention and emotional involvement in the topic. Yet the neuronal networks of our brains are not perfectly flexible. The very structure of our brain makes certain arithmetical concepts easier to "digest" than others.

I hope that the views I am defending here will eventually lead to improvements in teaching mathematics. A good curriculum would take into account the assets and limits of the learner's cerebral structure. To optimize the learning experiences of our children, we should consider what impact education and brain maturation have on the organization of mental representations. Obviously, we are still far from understanding to what extent learning can modify our brain machinery. The little that we already know could be of some use, however. The fascinating results that cognitive scientists have accumulated for the last twenty years on how our brain does math have not, until now, been made public and allowed to percolate through to the world of education. I would be delighted if this book served as a catalyst for improved communication between the cognitive and education sciences.

This book will take you on a tour of arithmetic as seen from the eyes of a biologist, but without neglecting its cultural components. In Chapters 1 and 2, through an initial visit of animals and human infant's abilities for arithmetic, I shall try to convince you that our mathematical abilities are not without biological precursors. Indeed, in Chapter 3 we shall find many traces of the animal mode of processing numbers still at work in adult human behavior. In Chapters 4 and 5, by observing how children learn to count and to calculate, we shall then attempt to understand how this initial approximate system can be overcome, and the difficulties that the acquisition of advanced mathematics raises for our primate brain. This will be a good occasion to investigate current methods of mathematical teaching and to examine the extent to which they have naturally adapted to our mental architecture. In Chapter 6 we shall also try to sort out the characteristics that distinguish a young Einstein or a calculating prodigy from the rest of us. In Chapters 7 and 8, finally, our number hunt will end up in the fissures of the cerebral cortex, where the neuronal circuits that support calculation are located, and from which, alas, they can be dislodged by a lesion or a vascular accident, thus depriving otherwise normal persons of their number sense.

Part I

Our Numerical Heritage

Chapter 1

Talented and Gifted Animals

One stone,
two houses
three ruins
four gravediggers,
one garden,
some flowers

One raccoon.
Jacques Prévert, *Inventaire*

Books on natural history have recounted the following anecdote since the eighteenth century:

> A nobleman wanted to shoot down a crow that had built its nest atop a tower on his domain. However, whenever he approached the tower, the bird flew out of gun range, and waited until the man departed. As soon as he left, it returned to its nest. The man decided to ask a neighbor for help. The two hunters entered the tower together, and later only one of them came out. But the crow did not fall into this trap and carefully waited for the second man to come out before returning. Neither did three, then four, then five men fool the clever bird. Each time, the crow would wait until all the hunters had departed. Eventually, the hunters came as a party of six. When five of them had left the tower, the bird, not so numerate after all, confidently came back, and was shot down by the sixth hunter.

Is this anecdote authentic? Nobody knows. It is not even clear that it has anything to do with numerical competence: For all we know, the bird could have memorized the visual appearance of each hunter rather than their number. Nevertheless, I decided to highlight it because it provides a splendid illustration of many aspects of animal arithmetic that are the subject of this chapter. First, in many tightly controlled experiments, birds and many other animal species appear to be able to perceive numerical quantities without requiring special training. Second, this perception is not perfectly accurate, and its accuracy decreases with increasingly larger numbers; hence the bird confounding 5 and 6. Finally, and more facetiously, the anecdote shows how the forces of Darwinian selection also apply to the arithmetical domain. If the bird had been able to count up to 6, it would perhaps never have been shot! In numerous species, estimating the number and ferocity of predators, quantifying and comparing the return of two sources of food, are matters of life and death. Such evolutionary arguments should help make sense of the many scientific experiments that have revealed sophisticated procedures for numerical calculation in animals.

A Horse Named Hans

At the beginning of this century, a horse named Hans made it to the headlines of German newspapers. His master, Wilhelm von Osten, was no ordinary circus animal trainer. Rather, he was a passionate man who, under the influence of Darwin's ideas, had set out to demonstrate the extent of animal intelligence. He wound up spending more than a decade teaching his horse arithmetic, reading, and music. Although the results were slow to come, they eventually exceeded all his expectations. The horse seemed gifted with a superior intelligence. It could apparently solve arithmetical problems and even spell out words!

Demonstrations of Clever Hans's abilities often took place in von Osten's yard. The public would form a half-circle around the animal and suggest an arithmetical question to the trainer—for instance, "How much is 5 plus 3?" Von Osten would then present the animal with five objects aligned on a table, and with three other objects on another. After examining the "problem," the horse responded by knocking on the ground with its hoof the number of times equal to the total of the addition. However, Hans's mathematical abilities far exceeded this simple feat. Some arithmetical problems were spoken aloud by the public, or were written in digital notation on a blackboard, and Hans could solve them just as easily (Figure 1.1). The horse could also add two fractions such as ⅖ and ½ and give the answer 9/10 by striking nine times, then ten times with its hoof. It was even said that to the question of determining the divisors of 28, Hans came out very appropriately with the answers 2, 4, 7, 14, and 28. Obviously, Hans's number

Figure 1.1. *Clever Hans" and his master Wilhelm von Osten strike a pose in front of an impressive array of arithmetic problems. The larger blackboard shows the numerical coding the horse used to spell words. (Copyright © Bildarchiv Preussicher Kulturbesitz.)*

knowledge surpassed by far what a school teacher would expect today of a reasonably bright pupil!

In September 1904, a committee of experts, among whom figured the eminent German psychologist Carl Stumpf, concluded after an extensive investigation that Hans's feats were real and not a result of cheating. This generous conclusion, however, did not satisfy Oskar Pfungst, one of Stumpf's own students. With von Osten's help—the master was fully convinced of his prodigy's superior intelligence—he began a systematic study of the horse's abilities. Pfungst's experiments, even by today's standards, remain a model of rigor and inventiveness. His working hypothesis was that the horse could not but be totally inept in mathematics. Therefore, it had to be the master himself, or someone in the public, who knew the answer and sent the animal a hidden signal when the target number of strokes had been reached, thus commanding the animal to stop knocking with its hoof.

To prove this, Pfungst invented a way of dissociating Hans's knowledge of a problem from what its master knew. He used a procedure that differed only slightly from the one described above. The master watched carefully as a simple addition was written in large printed characters on a panel. The panel was then oriented toward the horse in such a way that only it could see the problem and answer it. However, on some trials, Pfungst surreptitiously modified the addition before showing it to the horse. For instance, the master could see 6+2 whereas in fact the horse was trying to solve 6+3.

The results of this experiment, and of a series of follow-up controls, were clear-cut. Whenever the master knew the correct response, Hans got the right answer. When, on the contrary, the master was not aware of the solution, the horse failed. Moreover, it often produced an error that matched the numerical result expected by its master. Obviously, it was von Osten himself, rather than Hans, who was finding the solution to the various arithmetical problems. But how then did the horse know how to respond? Pfungst eventually deduced that Hans's truly amazing ability lay in detecting minuscule movements of its master's head or eyebrows that invariably announced the time to stop the series of knocks. In fact, Pfungst never doubted that the trainer was sincere. He believed that the signals were completely unconscious and involuntary. Even when von Osten was absent, the horse continued to respond correctly: Apparently, it detected the buildup of tension in the public as the expected number of hoof strokes was attained. Pfungst himself could never eliminate all forms of involuntary communication with the animal, even after he discovered the exact nature of the body clues it used.

Pfungst's experiments largely discredited demonstrations of "animal intelligence" and the competence of self-proclaimed experts such as Stumpf who had blindly subscribed to them. Indeed the Clever Hans phenomenon is still taught in psychology classes today. It remains a symbol of the pernicious influence that experimenter expectations and interventions, however small, may have on the outcome of any psychological experiment, with humans or with animals. Historically, Hans's story has played a crucial role in shaping the critical minds of psychologists and ethologists. It has drawn attention to the necessity for a rigorous experimental design. Since an essentially invisible stimulation as brief as the blink of an eye can influence the performance of animals, a well-designed experiment has to be devoid, from the start, of any possible source of errors. This lesson was particularly well-received by behaviorists, such as B. F. Skinner, who dedicated a large amount of work to the development of rigorous experimental paradigms for the study of animal behavior.

Unfortunately, Hans's exemplary case has also had more negative consequences on the development of psychological science. It has imposed an aura of suspicion onto the whole area of research on the representation of numbers in animals. Ironically, scientists now meet every single demonstration of numerical competence in animals with the same raised eyebrows that served as a cue to Hans! Such experiments are immediately associated, consciously or not, with Hans's story and are therefore suspected of a basic flaw in design, if not of downright forgery. This is an irrational prejudice, however. Pfungst's experiments showed only that Hans's numerical abilities were a fluke. By no means did they prove that it is impossible for an animal to understand some aspects of arithmetic. For a long time, however, the scientist's attitude was to systematically look for some experimental bias that might explain animal behavior without resorting to the hypothesis that animals have even an embryonic knowledge of calculation. For a while, even the most convincing results failed to convince anyone. Some researchers even preferred to attribute to animals mysterious abilities such as "rhythm discrimination" faculty, for instance, rather than admit that they could enumerate a collection of objects. In brief, the scientific community tended to throw out the baby with the bath water.

Before turning to some of the experiments that finally convinced all but the most skeptical of researchers, I would like to conclude Hans's story with a modern anecdote. Even today, the training of circus animals rests on methods rather similar to Hans's trick. If you ever see a show in which an animal adds numbers, spells words, or some surprising deed of this kind, you may safely bet that its behavior rests, like Hans's, on a hidden communication with its human trainer. Let me stress again that such communication need not be intentional. The trainer is often sincerely convinced of his pupil's gifts. A few years ago, I came upon an amusing article in a local Swiss newspaper. A journalist had visited the home of Gilles and Caroline P., whose poodle, named Poupette, seemed extraordinarily gifted in mathematics. Figure 1.2 shows Poupette's proud owner presenting his faithful and brilliant companion with a series of written digits that it was supposed to add. Poupette responded without ever making an error by tapping on its master's hand with its paw the exact number of times required, and then licking the hand after the correct count had been reached. According to its master, the canine prodigy had required only a brief training period, which led him to believe in reincarnation or some similar paranormal phenomenon. The journalist, however, wisely noted that the dog could react to subtle cues from the master's eyelids, or to some tiny motions of his hand when the correct count was reached. So this was indeed a case of reincarnation after all: the reincarnation of Clever Hans's stratagem, of which Poupette's story constituted, a century later, an astonishing replication.

Rat Accountants

Following the Hans episode, several renowned American laboratories developed research programs on animal mathematical abilities. Many such projects failed. A famous German ethologist named Otto Koehler, however, was more successful.

Figure 1.2. A modern canine "clever Hans": Poupette, the dog that could supposedly add digits.

One of his trained crows, Jacob, apparently learned to choose, among several containers, the one whose lid bore a fixed number of five points. Because the size, the shape, and the location of the points varied randomly from trial to trial, only an accurate perception of the number 5 could account for this performance. Nevertheless, the results achieved by Koehler's team had little impact, partly because most of their results were published only in German, and partly because Koehler failed to convince his colleagues that all possible sources of error, such as unintentional experimenter communication, olfactory cues or the like, had been excluded.

In the 1950s and 1960s, Francis Mechner, an animal psychologist at Columbia University, followed by John Platt and David Johnson at the University of Iowa, introduced a very convincing experimental paradigm that I shall schematically describe here. A rat that had been temporarily deprived of food was placed in a closed box with two levers, A and B. Lever B was connected to a mechanical device that delivered a small amount of food. However, this reward system did not work at once. The rat first had to repeatedly press lever A. Only after it had pressed for a fixed number of times *n* on lever A could it switch to lever B and get its deserved treat. If the rat switched too early to lever B, not only did it fail to get any food, but it received a penalty. On different experiments, the light could go off for a few seconds, or the counter was reset so that the rat had to start all over again with a new series of *n* presses on lever A.

How did rats behave in this rather unusual environment? They initially discovered, by trial and error, that food would appear when they pressed several times on lever A, and then once on lever B. Progressively, the number of times that they had to press was estimated more and more accurately. Eventually, at the end of the learning period, the rats behaved very rationally in relation to the number *n* that had been selected by the experimenter. The rats that had to press four times on lever A before lever B would deliver food, did press it about four times. Those that were placed in the situation where eight presses were required, waited until they had produced about eight squeezes, and so on (see Figure 1.3). Even when the requisite number was as high as twelve or sixteen, those clever rat accountants continued to keep their registers up to date!

Two details are worth mentioning. First, the rats often squeezed lever A a little more than the minimum required—five times instead of four, for instance. Again, this was an eminently rational strategy. Since they received a penalty for switching prematurely to lever B, the rats preferred to play it safe and press lever A once more rather than once less. Second, even after considerable training, the rats' behavior remained rather imprecise. Where the optimal strategy would have been to press lever A exactly four times, the rats often pressed it four, five, or

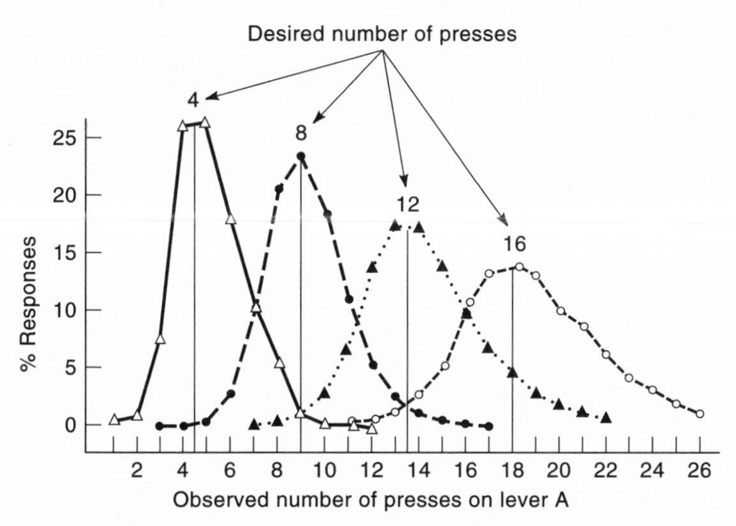

Figure 1.3. In an experiment by Mechner, a rat learns to press lever A a predetermined number of times before turning to a second lever B. The rat matches approximately the number selected by the experimenter, although its estimate becomes increasingly variable as the numbers get larger. (Adapted from Mechner 1958 by permission of the author and publisher; copyright © 1958 by the Society for the Experimental Analysis of Behavior.)

six times, and on some trials they even squeezed it three or seven times. Their behavior was definitely not "digital," and variation was considerable from trial to trial. Indeed, this variability increased in direct proportion to the target number that the rats estimated. When the target number of presses was four, the rats' responses ranged from three to seven presses, but when the target was sixteen, the responses went from twelve to twenty-four, thus covering a much larger interval. The rats appeared to be equipped with a rather imprecise estimation mechanism, quite different from our digital calculators.

At this stage, many of you are probably wondering whether I am not too liberal in attributing numerical competence to rats, and whether a simpler explanation of their behavior might not be found. Let me first remark that the Clever Hans effect cannot have any influence on this type of experiment because the rats are isolated in their cages and because all experimental events are controlled by an automated mechanical apparatus. However, is the rat really sensitive to the *number* of times the lever is pressed, or does it estimate the *time* elapsed since the beginning of a trial, or some other nonnumerical parameter? If the rat pressed at a regular rate, for instance once per second, then the above behavior might be fully explained by temporal rather than numerical estimation. While pressing on lever A, the rat would wait four, eight, twelve, or sixteen seconds, depending on the imposed schedule, before switching to lever B. This explanation might be considered simpler than the hypothesis that rats can count their movements—although, in fact, estimating duration and numbers are equally complex operations.

To refute such a temporal explanation, Francis Mechner and Laurence Guevrekian used a very simple control: They varied the degree of food deprivation imposed on the rats. When the rats are really hungry, and therefore eager to obtain their food reward as fast as possible, they press the levers much faster. Nevertheless, this increase in rate has absolutely no effect on the *number* of times they press the lever. The rats that are trained with a target number of four presses continue to produce between three and seven presses, while the rats trained to squeeze eight times continue to squeeze about eight times, and so on. Neither the average number of presses, nor the dispersion of the results, are modified with higher rates. Obviously, a numerical rather than a temporal parameter drives the rats' behavior.

A more recent experiment by Russell Church and Warren Meck, at Brown University, demonstrates that rats spontaneously pay as much attention to the number of events as to their duration. In Church and Meck's experiment, a loudspeaker placed in the rats' cage presented a sequence of tones. There were two possible sequences. Sequence A was made up of two tones and lasted a total of two seconds, whereas sequence B was made up of eight tones and lasted eight seconds. The rats had to discriminate between the two melodies. After each tune, two levers were inserted in the cage. To receive a food reward, the rats had to press the left lever if they had heard sequence A, and the right if they had heard sequence B (see Figure 1.4).

Several preliminary experiments had shown that rats placed in this situation rapidly learned to press the correct lever. Obviously, they could use two distinct parameters to distinguish A from B: the total duration of the sequence (two versus eight seconds) or the number of tones (two versus eight). Did rats pay attention to duration, number, or both? In order to find out, the experimenters presented some test sequences in which duration was fixed while number was varied, and others in which number was fixed while duration was varied. In the

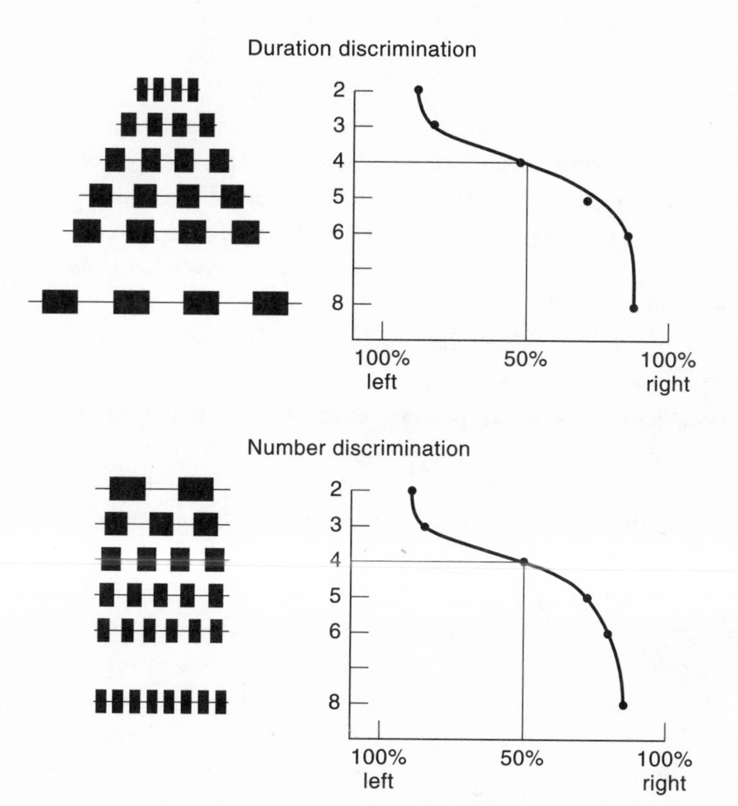

Figure 1.4. Meck and Church trained rats to press a lever on the left when they heard a short two-tone sequence, and a lever on the right when they heard a long eight-tone sequence. Subsequently, the rats generalized spontaneously: for equal numbers of sounds, they discriminated two-second sequences from eight-second sequences (top panel), and for an equal total duration, they discriminated two tones from eight tones (bottom panel). In both cases, four seems to be the "subjective middle" of 2 and 8, the point where rats cannot decide whether they should press right or left. (Adapted from Meck and Church 1983.)

first case, all sequences lasted four seconds, but were made up of from two to eight tones. In the second case, all sequences were made up of four tones, but duration extended from two to eight seconds. On all such test sequences, the rats always received a food reward, regardless of the lever they picked. In anthropocentric terms, the researchers were simply asking what these new stimuli sounded like to the rats, without letting the reward interfere with their decision. The experiment therefore measured the rats' ability to generalize previously learned behaviors to a novel situation.

The results are clear-cut. Rats generalized just as easily on duration as on number. When duration was fixed, they continued to press the left lever when they heard two tones, and the right lever when they heard eight tones. Conversely, when number was fixed, they pressed left for two-second sequences, and right for eight-second sequences. But what about intermediate values? Rats apparently reduced them to the closest stimulus that they had learned. Thus, the new three-tone sequence elicited the same response as the two-tone sequence used for training, while sequences with five or six tones were classified just as the original sequence of eight tones had been. Curiously, when the sequence comprised just four tones, the rats could not decide whether they should press left or right. For a rat, four appears to be the subjective midpoint between the numbers two and eight!

Keep in mind that the rats did not know during training that they would be tested subsequently with sequences that varied in duration or in number of tones. Hence, this experiment shows that when a rat listens to a melody, its brain simultaneously and spontaneously registers both the duration and the number of tones. It would be a serious mistake to think that because these experiments use conditioning, they somehow teach the rats how to count. On the contrary, rats appear on the scene with state-of-the-art hardware for visual, auditory, tactile, and numerical perception.Conditioning merely teaches the animal to associate perceptions that it has always experienced, such as representations of stimulus duration, color, or number, with novel actions such as pressing a lever. There is no reason to think that number is a complex parameter of the external world, one that is more abstract than other so-called objective or physical parameters such as color, position in space, or temporal duration. In fact, provided that an animal is equipped with the appropriate cerebral modules, computing the approximate number of objects in a set is probably no more difficult than perceiving their colors or their positions.

Indeed, we now know that rats and many other species spontaneously pay attention to numerical quantities of all kinds—actions, sounds, light flashes, food morsels. For instance researchers have proved that raccoons, when presented with several transparent boxes with grapes inside, can learn to systematically select those that contain three grapes and to neglect those that contain two or four. Likewise, rats have been conditioned to systematically take the fourth tunnel on the left in a maze, regardless of the spacing between consecutive tunnels. Other researchers have taught birds to pick the fifth seed that they find when visiting several interconnected cages. And pigeons can, under some circumstances, estimate the number of times they have pecked at a target and can discriminate, for instance, between forty-five and fifty pecks. As a final example, several animals, including rats, appear to remember the number of rewards and punishments that they have received in a given situation. An elegant experiment by E. J. Capaldi and Daniel Miller at Purdue University has even shown that when rats receive food rewards of two different kinds—say, raisins and cereals—they keep in mind three pieces of information at the same time: the number of raisins that they have eaten, the number of pieces of cereals, and the total number of food items. In brief, far from being an exceptional ability, arithmetic is quite common in the animal world. The advantages that it confers for survival are obvious. The rat that remembers that its hideout is the fourth to the left will move faster in the dark maze of tunnels that it calls home. The squirrel that notices that a branch bears two nuts, and neglects it for another one that bears three, will have more chances of making it safely through the winter.

How Abstract Are Animal Calculations?

When a rat presses a lever twice, hears two sounds, and eats two seeds, does it recognize that these events are all instances of the number "2"? Or can't it see the link between numbers that are perceived through different sensory modalities? The ability to generalize across different modalities of perception or action is an important component of what we call the *number concept*. Let us suppose, as an admittedly extreme case, that a child systematically utters the word "four" whenever he or she sees four objects, but randomly picks the words "three," "four," or "nine" when he or she hears four sounds or makes four jumps. Although performance is no doubt excellent with visual stimuli, we would be reluctant to grant the child knowledge of the concept of "4", because we consider possession of this concept to entail being able to apply it to many different multimodal situations. As a matter of fact, as soon as children have learned a number word, they can immediately use it to count their toy cars, the meows of their cat, or the misdemeanors of their little brother. What about rats? Is their numerical competence confined to certain sensory modalities, or is it abstract?

Unfortunately, any answer must remain tentative because few successful experiments have been done on multimodal generalization in animals. However, Russell Church and Warren Meck have shown that rats represent number as an abstract parameter that is not tied to a specific sensory modality, be it auditory or visual. They again placed rats in a cage with two levers, but this time stimulated them with visual as well as with auditory sequences. Initially, the rats were conditioned to press the left lever when they heard two tones, and the right lever when they heard four tones. Separately, they were also taught to associate two light flashes with the left lever, and four light flashes with the right lever. The issue was, how were these two learning experiences coded in the rat brain? Were they stored as two unrelated pieces of knowledge? Or had the rats learned an abstract rule such as "2 is left, and 4 is right"? To find out, the two researchers presented mixtures of sounds and light flashes on some trials. They were amazed to observe that when they presented a single tone synchronized with a flash, a total of two events, the rats immediately pressed the left lever. Conversely, when they presented a sequence of two tones synchronized with two light flashes, for a total of four events, the rats systematically pressed the right lever. The animals generalized their knowledge to an entirely novel situation. Their concepts of the numbers "2" and "4" were not linked to a low level of visual or auditory perception.

Consider how peculiar the rats' behavior was on trials with two tones synchronized with two light flashes. Remember that in the course of their training, the rats were always rewarded for pressing the left lever after hearing two tones, and likewise after seeing two flashes of light. Thus, both the auditory "two tones" stimulus and the visual "two flashes" stimulus were associated with pressing the left lever. Nevertheless, when these two stimuli were presented together, the rats pressed the lever that had been associated with the number 4! To better grasp the significance of this finding, compare it with a putative experiment in which rats are trained to press the left lever whenever they see a square (as opposed to a circle), and to respond left whenever they see the color red (as opposed to green). If the rats were presented with a red square—the combination of both stimuli—I bet that they would press even more decidedly on the left lever. Why are the numbers of tones and flashes grasped differently from shapes and colors? The experiment demonstrates that rats "know," to some extent, that numbers do not add up in the same way as shapes and colors. A square plus the color red makes a red square, but two tones plus two flashes do not evoke an even greater sensation of twoness. Rather, 2 plus 2 makes 4, and the rat brain seems to appreciate this fundamental law of arithmetic.

Perhaps the best example of abstract addition abilities in an animal comes from work done by Guy Woodruff and David Premack at the University of Pennsylvania. They set out to prove that a chimpanzee could do arithmetic with simple fractions. In their first experiment, the chimpanzee's task was simple: It was rewarded for selecting, among two objects, the one that was physically identical to a third one. For instance, when presented with a glass half-filled with a blue liquid, the animal had to point toward the identical glass when presented next to another glass that was filled up to three-quarters of its volume. The chimp immediately mastered this simple physical matching task. Then the decision was progressively made more abstract. The chimp might be shown a half-full glass again, but now the options were either half an apple or three-quarters of an apple. Physically speaking, both alternatives differed widely from the sample stimulus; yet the chimpanzee consistently selected the half apple, apparently basing its responses on the conceptual similarity between half a glass and half an apple. Fractions of one-quarter, one-half, and three-quarters were tested with similar success: The animal knew that one-quarter of a pie is to a whole pie as one-quarter of a glass of milk is to a full glass of milk.

In their last experiment, Woodruff and Premack showed that chimpanzees could even mentally combine two such fractions: When the sample stimulus was made of one-quarter apple and one-half glass, and the choice was between one full disc or three-quarters disc, the animals chose the latter more often than chance alone would predict. They were obviously performing an internal computation not unlike the addition of two fractions: $^1/_4 + ^1/_2 = ^3/_4$. Presumably, they did not use sophisticated symbolic calculation algorithms as we would. But they clearly had an intuitive grasp of how these proportions should combine.

A final anecdote concerning Woodruff and Premack's work: Though the manuscript reporting their work was initially titled "Primitive mathematical concepts in the chimpanzee: proportionality and numerosity," an editorial error made it appear in the pages of the scientific journal *Nature* under the heading "*Primative* mathematical concepts . . . "! Involuntary as it was, this alteration was not so improper. For primitive, indeed, the animal's ability was not. And if "primative" was taken to mean "specific to primates," then the neologism seemed very appropriate here because such an abstract ability to add fractions has not been observed in any other species so far.

Addition, however, is not the only numerical operation in the animal repertoire. The ability to compare two numerical quantities is an even more fundamental ability, and indeed it is widespread among animals. Show a chimpanzee two trays on which you have placed several bits of chocolate. On the first tray, two piles of chocolate chips are visible, one with four pieces, and the other with three pieces. The second tray contains a pile with five pieces of chocolate and, separate from it, a single piece. Leave the animal enough time to watch the situation carefully before letting it choose one tray and eat its content. Which tray do you think that it will pick? Most of the time, without training, the chimpanzee selects the tray with the largest total number of chocolate chips (see Figure 1.5).

Figure 1.5. A chimpanzee spontaneously selects the pair of trays with the greater total *number of chocolate bits, revealing its inborn ability to add and compare approximate numerosities. (Reprinted from Rumbaugh et al. 1987.)*

Hence, the greedy primate must spontaneously compute the total of the first tray (4+3=7), then the total of the second tray (5+1=6), and finally it must reckon that 7 is larger than 6 and that it is therefore advantageous to choose the first tray. If the chimp could not do the additions, but was content with choosing the tray with the largest single pile of chocolates, it should have been wrong in this particular example because while the pile with five chips on the second tray exceeds each of the piles on the first tray, the total amount of chips on the first tray is larger. Clearly, the two additions and the final comparison operation are all required for success.

Although chimps perform remarkably well in selecting the larger of two numbers, their performance is not devoid of errors. As is frequently the case, the nature of these errors provides important cues about the nature of the mental representation employed. When the two quantities are quite different, such as 2 and 6, chimpanzees hardly ever fail: They always select the larger. As the quantities become closer, however, performance systematically decreases. When the two quantities differ by only one unit, only 70% of the chimp's choices are correct. This systematic dependency of error rate on the numerical separation between the items is called the *distance effect*. It is also accompanied by a *magnitude effect*. For equal numerical distances, performance decreases as the numbers to be compared become larger. Chimpanzees have no difficulty in determining that 2 is larger than 1, even though these two quantities differ only by one unit. However, they fail increasingly more often as one moves to larger numbers such as 2 versus 3, 3 versus 4, and so on. Similar distance and magnitude effects have been observed in a great variety of tasks and in many species, including pigeons, rats, dolphins, and apes. No animals seem able to escape these laws of behavior—including, as we shall see later, *Homo sapiens*.

Why are these effects of distance and magnitude important? Because they demonstrate, once again, that animals do not possess a digital or discrete representation of numbers. Only the first few numbers—1, 2, and 3—can be discriminated with high accuracy. As soon as one advances toward larger quantities, fuzziness increases. The variability in the internal representation of numbers grows in direct proportion to the quantity represented. This is why, when numbers get large, an animal has problems distinguishing number n from its successor $n+1$. One should not conclude, however, that large numbers are out of reach of the rat or pigeon brain. In fact, when numerical distance is sufficiently large, animals can successfully discriminate and compare very large numbers, on the order of 45 versus 50. Their imprecision simply leaves them blind to the finesses of arithmetic, such as the difference between 49 and 50.

Within the limits set by this internal imprecision, we have seen through numerous examples, animals possess functional mathematical tools. They can add two quantities and spontaneously choose the larger of two sets. Should we really be that surprised? Let us first try to think whether the outcome of these experiments could possibly have been any different. When a hungry dog is offered a choice between a full dish and a half-full one of the same food, doesn't it spontaneously pick the larger meal? Acting otherwise would be devastatingly irrational. Choosing the larger of two amounts of food is probably one of the preconditions for the survival of any living organism. Evolution has been able to conceive such complex strategies for food gathering, storing, and predation, that it should not be astonishing that an operation as simple as the comparison of two quantities is available to so many species. It is even likely that a mental comparison algorithm was discovered early on, and perhaps even reinvented several times in the course of evolution. Even the most elementary of organisms, after all, are confronted with a never-ending search for the best environment with the most food, the fewest predators, the most partners of the opposite sex, and so on. One must optimize in order to survive, and compare in order to optimize.

We still have to understand, however, by what neural mechanisms such calculations and comparisons are carried out. Are there minicalculators in the brains of birds, rats, and primates? How do they work?

The Accumulator Metaphor

How can a rat know that 2 plus 2 makes 4? How can a pigeon compare forty-five pecks with fifty? I know by experience that these results are often met with disbelief, laughter, or even exasperation—especially when the audience is composed of professors of mathematics! Our Western societies, ever since Euclid and Pythagoras, have placed mathematics at the pinnacle of human achievements. We view it as a supreme skill that either requires painful education or comes as an innate gift. In many a philosopher's mind, the human ability for mathematics derives from our competence for language, so that it is inconceivable that an animal without language can count, much less calculate with numbers.

In this context, the observations about animal behavior that I have just described are in danger of being simply disregarded, as often happens with unexpected or seemingly aberrant scientific results. Without a theoretical framework to support them, they might appear as isolated findings, peculiar indeed, but eventually inconclusive and certainly not sufficient to question the equation "mathematics = language." To sort out such phenomena, we need a theory that explains, quite simply, how it is possible to count without words.

Fortunately, such a theory exists. In fact, we all know of mechanical devices whose performances are not so different from those of rats. All cars, for instance, are equipped with a counting mechanism that keeps a record of the number of miles that have accumulated since the vehicle was first put in circulation. In its simplest version, this "counter" is just a cog wheel that advances by one notch for each additional mile. At least in principle, this example shows how a simple mechanical device may keep a record of an accumulated quantity. Why could a biological system not incorporate similar principles of counting?

The car counter is an imperfect example because it uses digital notation, a symbolic system that is most probably specific to humans. In order to account for the arithmetical abilities of animals, we should look for an even simpler metaphor. Imagine Robinson Crusoe, on his desert island, alone and helpless. For the sake of argument, let us even imagine that a blow to the head has deprived him of any language, leaving him unable to use number words for counting or calculation. How could Robinson build an approximate calculator using only the makeshift means available to him? This is actually easier than it would seem. Suppose that Robinson has discovered a spring in the vicinity. He digs an accumulator tank in the trunk of a large tree, and places this accumulator next to the stream, so that water does not flow directly into it but can be temporarily diverted by using a small bamboo pipe. With this rudimentary device, of which the accumulator is the central component, Robinson will be able to count, add, and compare approximate numerical magnitudes. In essence, the accumulator enables him to master arithmetic as well as a rat or a pigeon does.

Suppose that a canoe loaded with cannibals approaches Robison's island. How can Robinson, who is following this scene with a telescope, keep a record of the number of attackers using his calculator? First, he would have to empty the accumulator. Then, each time a cannibal landed, Robinson would briefly divert some water from the spring into the accumulator. Furthermore, he does this so that it always takes a fixed amount of time and that the water flow remains constant throughout. Thus, for each attacker to be counted, a more or less fixed amount of water flows into the accumulator. In the end, the water level in the accumulator will be equal to n times the amount of water diverted at each step. This final water level may then serve as an approximate representation of the number n of cannibals who have landed. This is because it depends only on the number of events that have been counted. All other parameters, such as the duration of each event, the time interval between them, and so on, have no influence on it. The final level of water in the accumulator is thus completely equivalent to number.

By marking the level reached by water in the accumulator, Robinson can keep a record of how many people have landed, and he may use this number in later calculations. The next day, for instance, a second canoe approaches. To estimate the total number of attackers, Robinson first fills the accumulator up to the level of the preceding day's marker, and then adds a fixed amount of water for each newcomer, just as he did previously. The new water level, after this operation is completed, will represent the result of the addition of attackers in the first canoe and in the second. Robinson can keep a permanent record of this computation by carving a different mark on the accumulator.

The day after, a few savages leave the island. To evaluate their number, Robinson empties his accumulator and repeats the above procedure, adding some water for each departing cannibal. He realizes that the final water level, which represents the number of people who have left, is much lower than the previous day's mark. By comparing the two water levels, Robinson reaches the worrisome conclusion that, in all likelihood, the number of natives that have left is smaller than the number of natives that have arrived in the past two days. In brief, Robinson, using his rudimentary device, can count, compute simple additions, and compare the results of his calculations, just like the animals in the above experiments.

A clear drawback of the accumulator is that numbers, although they form a discrete set, are represented by a continuous variable: water level. Given that all physical systems are inherently variable, the same number may be represented, at different times, by different amounts of water in the accumulator. Let us suppose, for instance, that water flow is not perfectly constant and varies randomly by between 4 and 6 liters per second, with a mean of 5 liters per second. If Robinson diverts water for two-tenths of a second into the accumulator, one liter on average will be transferred. However, this quantity will vary from 0.8 to 1.2 liters. Thus, if five items are counted, the final water level will vary by between 4 and 6 liters. Given that the very same levels could have been reached if four or six items had been counted, Robinson's calculator is unable to reliably discriminate the numbers 4, 5, and 6. If six cannibals land, and later only five depart, Robinson is in danger of failing to notice that one of them is missing. This, by the way, is exactly the situation that confronted the crow in the anecdote I mentioned at the beginning of this chapter! Robinson clearly will be better able to discriminate numbers that are more different; this is the distance effect. This effect will be exacerbated as the numbers become larger, thus reproducing the magnitude effect that also characterizes animal behavior.

One might object that the imaginary Robinson I am describing is not particularly clever. What prevents him from using marbles instead of imprecise amounts of water? Dropping in a bowl a single marble for each counted item would provide him with a discrete and precise representation of their number. In this manner, he would avoid errors even in the most complex of subtractions. But Robinson's machine is used here only as a metaphor for the animal brain. The nervous system—at least the one that rats and pigeons possess—does not seem to be able to count using discrete tokens. It is fundamentally imprecise and seems unable to precisely keep track of the items that it counts; hence its increasing variance for larger and larger numbers.

Although the accumulator model is described here in a very informal manner, it is actually a rigorous mathematical model, the equations of which accurately predict variations in animal behavior as a function of number size and numerical distance. The accumulator metaphor thus helps us to understand why rat behavior is so variable from one trial to the next. Even after considerable training, a rat seems unable to press exactly four times on a lever, but it can press four, five, or six times on different trials. I believe that this is due to a fundamental inability to represent numbers 4, 5, and 6 in a discrete and individualized format, as we do. To a rat, numbers are just approximate magnitudes, variable from time to time, and as fleeting and elusive as the duration of sounds or the saturation of colors. Even when an identical sequence of sounds is played twice, rats probably do not perceive the exact same number of sounds, but only the fluctuating level of an internal accumulator.

Of course, the accumulator is nothing more than a vivid metaphor that merely illustrates how a simple physical device can mimic, in considerable detail, experiments on animal arithmetic. There are no taps and recipients in the brains of rats and pigeons. Would it be possible, however, to identify, within the cerebrum, neuronal systems that might occupy a function similar to the components in the accumulator model? This is a completely open question. Currently, scientists are merely beginning to understand how certain parameters are modified by various pharmacological substances. Injecting rats with metamphetamine, for instance, seems to accelerate the internal counter. The rats injected with this substance respond to a sequence of four sounds as if they had been five or six. It is as if the flow of water to the accumulator were accelerated by metamphetamine. For each item counted, an amount of water larger than usual reaches the accumulator, thus making the final water level too great. This is how a 4, in the input, may end up looking like a 6 at the output. We still have little knowledge, however, of the brain regions in which metamphetamine produces its accelerating effect. Cerebral circuitry is far from having revealed all its secrets.

Number-Detecting Neurons?

Although the cerebral circuits for number processing remain largely unknown, neural network simulations can be used to speculate on what their organization may be like. Neural network models are algorithms that run on a conventional digital computer, but emulate the kind of computations that may go on in real brain circuits. Of course, the simulations are always vastly simplified when compared to the overarching complexity of real networks of neurons. In most computer models, each neuron is reduced to a digital unit with an output level of activation varying between 0 and 1. Active units excite or inhibit their neighbors as well as more distant units via connections with a variable weight, which are analogous to the synapses that connect real neurons. At each step, each simulated unit sums up the inputs it receives from other units and switches on or off depending on whether the sum exceeds a given threshold. The analogy to a real nerve cell is crude, but one crucial property is preserved: the fact that a great many simple computations take place at the same time in several neurons distributed within multiple circuits. Most neurobiologists believe that such massive parallel processing is the key property that enables brains to perform complex computations in a short time using relatively slow and unreliable biological hardware.

Can parallel neuronal processing be used to process numbers? With Jean-Pierre Changeux, a neurobiologist at the Pasteur Institute in Paris, I have proposed a tentative neural network simulation of how animals extract numbers from their environment quickly and in parallel. Our model addresses a simple problem that rats and pigeons routinely solve: given an input retina on which objects of various sizes are displayed, and given a cochlea on which tones of various frequencies are played, can a network of simulated neurons compute the total number of visual and auditory objects? According to the accumulator model, this number can be computed by adding to an internal accumulator a fixed quantity for each input item. The challenge is to do this with networks of simulated nerve cells, and to achieve a representation of number that is independent of the size and location of visual objects as well as of the time of presentation of auditory tones.

We solved the problem by first designing a circuit that normalizes the visual input with respect to size. This network detects the locations occupied by objects on the retina, and allocates to each object, regardless of size and shape, an approximately constant number of active neurons on a location map. This normalization step is crucial because it allows the network to count each object as "one," regardless of size. As we shall see below, in mammals this operation may be achieved by circuits of the posterior parietal cortex, which are known to compute a representation of object location without taking exact shape and size into account.

In our simulation, a similar operation is also performed for auditory stimuli. Regardless of the time intervals at which they are received, auditory inputs are accumulated in a single memory store. Once these normalizations for size, shape, and time of presentation have been accomplished, it is easy to estimate number—one simply has to evaluate the total neuronal activity in the normalized visual map and in the auditory memory store. This total is equivalent to the final water level in the accumulator, and it provides a reasonably reliable estimate of number. In our simulation, the summation operation is taken care of by an array of units that pool activations from all the underlying visual and auditory units. Under certain conditions, these output units fire only when the total activity they receive falls within a predefined interval that varies from one neuron to the next. Each of these simulated neurons therefore works as a number detector that reacts only when a certain approximate number of objects is seen (Figure 1.6). One unit in the network, for instance, responds optimally when presented with four objects—be they, for instance, four visual blobs, four sounds, or two blobs and two sounds. The same unit reacts infrequently when presented with three or five objects, and not at all in all other cases. It therefore works as an abstract detector of number 4. The entire number line can be covered by such detectors, each tuned to a different approximate number, with the precision of tuning decreasing as one moves to increasingly larger numbers. Because the simulated neurons process all visual and auditory inputs simultaneously, the array of number detectors responds very quickly—it can estimate the cardinal of a set of four objects in parallel over the entire retina, without having to orient in turn toward each item as we do when we count.

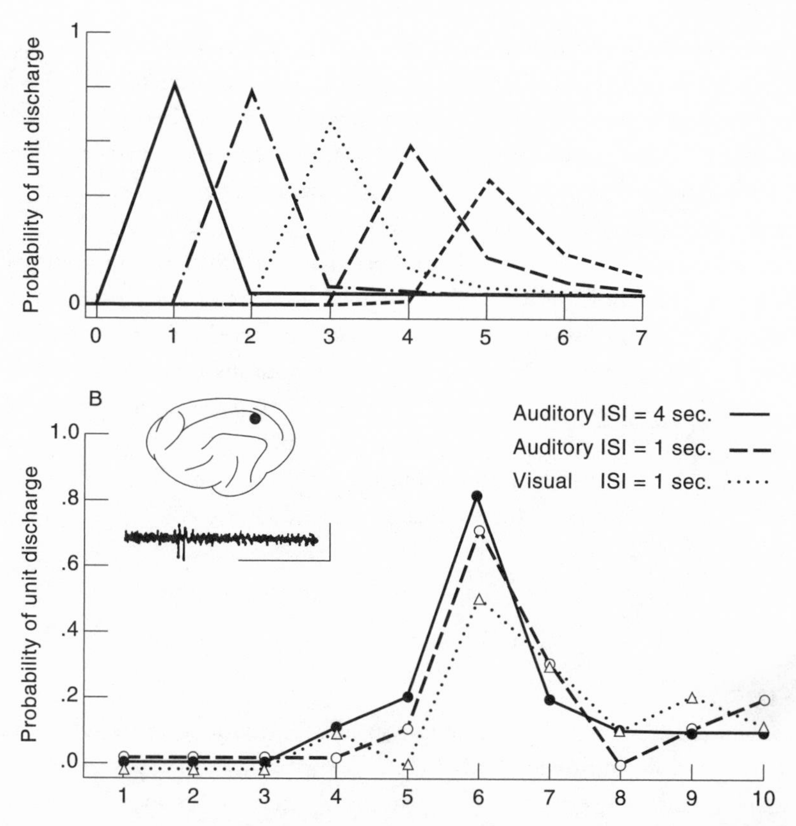

Figure 1.6. A computer-simulated neural network incorporates "numerosity detectors" that respond preferentially to a specific number of input items (top panel). Each curve shows the response of a given unit to different numbers of items. Note the decreasing selectivity of responses as input numerosity increases. In 1970, Thompson and his colleagues recorded similar "number-coding" neurons in the association cortex of anesthetized cats (bottom panel). The neuron illustrated here responds preferentially to six consecutive events, either six flashes of light one second apart, or six tones one or four seconds apart. (Top, adapted from Dehaene and Changeux 1993; bottom, Thompson et al. 1970. Copyright © 1970 by American Association for the Advancement of Science).

Astonishingly, the number-detecting neurons that the model predicts seem to have been identified at least once in an animal brain. In the 1960s, Richard Thompson, a neuroscientist at the University of California at Irvine, recorded the activity of single neurons in the cortex of cats while the animals were presented with series of tones or of light flashes. Some cells fired only after a certain number of events. One neuron, for instance, reacted after six events of any kind, regardless of whether this was six flashes of light, six brief tones, or six longer tones. Sensory modality did not seem to matter: The neuron apparently cared only about number. It did not respond in a discrete all-or-none manner either, unlike a digital computer. Rather, its activation level grew after the fifth item, reached a peak for the sixth, and decreased for larger numbers of items, a response profile quite similar to that of the simulated neurons in our model. Several similar cells, each tuned to a different number, were recorded in a small area of the cat's cortex.

Thus, there might well be a specialized brain area, equivalent to Robinson's accumulator, in the animal brain. Unfortunately, Thompson's study, published in the prestigious scientific journal *Science* in 1970, did not receive further attention. We still have no idea whether the number-detecting neurons are connected in the way our model predicts, or whether cats' brains extract number using some other method. The final word on this story will no doubt belong to those neurophysiologists who will dare to continue the quest for the neuronal bases of animal arithmetic using modern neuronal recording tools.

Fuzzy Counting

Whatever its exact neuronal implementation, if the accumulator model is correct, two conclusions must necessarily follow. First, animals can count, since they are able to increase an internal counter each time a external event occurs. Second, they do not count exactly as we do. Their representation of numbers, contrary to ours, is a fuzzy one.

When we count, we use a precise sequence of number words, leaving no room for errors to creep in. Each item counted corresponds to a move of one step forward in the number sequence. Not so for rats. Their numbers are the floating levels of an analogical accumulator. When a rat adds one unit to its running total, the operation bears only a vague resemblance to the logical rigorousness of our "+1." It is more like adding a bucket of water to Robinson's accumulator. The rat's condition is somewhat reminiscent of Alice's arithmetical embarrassment in *Through the Looking Glass*:

> "Can you do Addition?" the White Queen asked. "What's one and one and one and one and one and one and one and one and one and one?"
>
> "I don't know," said Alice. "I lost count."
>
> "She can't do Addition," the Red Queen interrupted.

Presumably, although she lacked enough time to count verbally, Alice would have been able to estimate the total to within a few units. Likewise, rats have to resort to approximate counting without words or digital symbols. The difference with our verbal counting is so enormous that we should perhaps not talk about "number" in animals at all, because by number we often imply a discrete symbol. This is why scientists, when they describe perception of numerical quantities, speak of "numerosity" or "numerousness" rather than number. The accumulator enables animals to estimate how numerous some events are, but does not allow them to compute their exact number. The animal mind can retain only fuzzy numbers.

Is it really impossible to teach animals a symbolic notation for numbers? Couldn't we teach them to recognize a discrete set of numerical labels similar to our digits and number words, and then inculcate to them that these labels refer to precise quantities? In fact, several such experiments have met with mitigated success. In the 1980s, a Japanese researcher, Tetsuro Matsuzawa, taught a chimpanzee named Ai the use of arbitrary signs to describe sets of objects (Figure 1.7). The small drawings that played the role of words occupied the cells of a computerized pad. The chimp could press any cells that he chose in order to describe what it saw. After a long training period, Ai learned to use fourteen object symbols, eleven color symbols and, most important for us, the first six Arabic numerals. When it was shown three red pencils, for instance, the chimp first pointed toward a square symbol adorned with a black diamond, which conventionally meant "pencil," then toward a diamond crossed by a horizontal bar ("red"), and finally toward the written digit "3."

This sequence of gestures may have been only some elaborate form of rote motor reflex. However, Matsuzawa showed that the drawings did, to some extent, function like words that could, through their combinations alone,

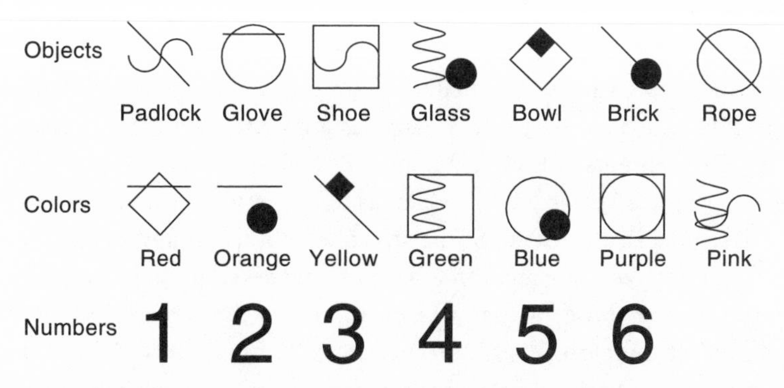

Figure 1.7. The Japanese primatologist Matsuzawa taught his chimpanzee Ai a vocabulary of visual signs, of which only a small subset is presented here. Ai could thus report the identity, the color and the numerosity of small sets of objects. (After Matsuzawa, 1985; copyright © 1985 by Macmillan Magazines Ltd.)

describe novel situations. If, for instance, the chimpanzee was taught a new symbol for "toothbrush," it was partially able to apply it to novel contexts such as "five green toothbrushes" or "two yellow toothbrushes." Still, this ability to generalize remained fraught with frequent errors.

Since 1985, when Matsuzawa first reported his results, his chimpanzee Ai has made constant progress in arithmetic. It now knows the first nine digits and can enumerate sets with 95% accuracy. Recordings of his response times suggest that, like a human, Ai uses serial counting for numbers greater than 3 or 4. It has also learned to order the digits according to their magnitude—although again, it took years to establish this novel competence.

Since Matsuzawa's early experiments, the learning of numerical labels has been replicated in several chimpanzees from at least three different primate training centers. Similar abilities have even been demonstrated in species much more distant from us. Dolphins have been trained to associate arbitrary objects with precise numbers of fish. After about two thousand trials, they were able to select, among two objects, the one that was associated with the larger amount of fish. Irene Pepperberg at the University of Arizona has taught her parrot Alex a large vocabulary of English words, among which are the first few number words. Experiments with Alex are quite remarkable in that signs or plastic tokens are unnecessary: More or less standard English can be used to formulate oral questions, which the animal immediately answers by uttering recognizable words! When it is presented with an array of objects comprising, for instance, green keys, red keys, green toys, and red toys, Alex can answer questions as complex as "How many red keys?" Naturally, his training took a long time—almost twenty years. However, the results clearly prove that numerosity labeling is not exclusively a mammal's privilege.

In more recent work, chimpanzees have been shown to be partially able to calculate using numerical symbols. Sarah Boysen, for instance, taught her chimpanzee named Sheba to perform simple numerical additions and comparisons. She started by teaching Sheba the quantities associated with the Arabic digits 0 through 9. Experiments of this kind require unfailing patience. For two years, the animal was progressively exposed to increasingly complex tasks. At first, it simply had to place one biscuit in each of the six squares of a checkerboard. It was then shown sets of between one and three biscuits, and asked to select, among several cards, the one that bore as many black marks as there were biscuits on the checkerboard. It therefore learned to match a set of biscuits with a set of marks by focusing only on their numerosity. In a third stage, the cards with marks were progressively replaced with the corresponding Arabic digits. The chimp therefore learned to recognize the digits 1, 2, and 3, and to point to the appropriate digit when it saw the corresponding number of biscuits. Finally, in the last stage, Sarah

Boysen taught her protégé the converse: It had to choose, among several sets of objects, the one whose numerosity matched a given Arabic digit.

Using similar strategies, the knowledge of the animal was progressively extended to the entire set of digits, from 0 through 9. At the end of this training period, Sheba could fluently move back and forth between a digit and the corresponding quantity. This can be considered as the essence of symbolic knowledge. A symbol, beyond its arbitrary shape, refers to a covert meaning. Symbol comprehension implies accessing this meaning from shape alone, while symbol production requires recovering the arbitrary shape from knowledge of the intended meaning. Obviously, the chimpanzee Sheba had managed, through long and painful training, to master both of these transformations.

An important property of human symbols, though, is that they can be combined into sentences whose meaning derives from the meaning of the constituent words. Mathematical symbols, for instance, can be combined to express equations such as 2+2=4. Could Sheba also combine multiple digits into a symbolic calculation? To find out, Boysen designed a symbolic addition task. She hid oranges at several places in Sheba's cage—for instance, two oranges under a table and three in a box. The chimpanzee first explored the various places where the oranges might be hidden. Sheba then came back to the starting point and was supposed to pick, among several Arabic digits, the one that matched the total number of oranges found. From the very first trial, the animal succeeded. A symbolic version of it was immediately tried. This time, as it wandered through the cage, the animal did not discover oranges, but Arabic digits such as digit 2 under the table and digit 4 in the box. Again, right from the start, the chimpanzee was regularly able to report, when its exploration was over, the total of the digits that it had seen (2+4=6). This implied that it could recognize each of the digits, associate them mentally with quantities, figure out the result of adding together all these quantities, and finally retrieve the visual appearance of the corresponding digit. Never had an animal come any closer to the symbolic calculation abilities exhibited by humankind.

Even species far less clever that the chimpanzee can learn to perform elementary mental operations with numerical symbols. For instance, two macaques named Abel and Baker, trained by David Washburn and Duane Rumbaugh at Georgia State University, have shown remarkable abilities for comparing the numerical quantities conveyed by Arabic digits. Pairs of Arabic digits such as "2 4" appeared on a computer screen. Using the joystick, the animal could choose one digit. An automated dispenser then delivered a corresponding number of fruit candies, a delicacy that primates are particularly fond of. If the animal chose digit 4, it could savor four candies, whereas if it selected digit 2, it would only get two. The drive toward choosing the larger digit was therefore quite important. Indeed, the task was rather similar to the above-described comparison task, except that the animal was not directly confronted with food, but only with a symbolic representation of its amount using Arabic digits. It had to retrieve from memory the meaning of the digit symbols—namely, the quantity with which they were associated.

I should mention that Abel and Baker, unlike Sheba, had not received any training with Arabic digits before the test started. This is why they needed several hundred trials to learn to choose the larger digit with some regularity. Sheba, who already knew the quantity associated with digits, answered correctly on the very first trial of a similar number comparison task. After training, Abel and Baker also succeeded very well. They made no mistakes at all when the digits were sufficiently distant, but they failed up to 30% of the time when the digits differed by only one unit. We recognize here the now-familiar distance effect, which reveals a tendency to confound quantities that are numerically close.

Following this performance with digit pairs, Abel and Baker went on successfully to triplets, quadruplets, and even quintuplets of digits between 1 and 9. Clearly, the animals had not learned the answers to all possible pairs of digits by rote. Even when they were presented with a new randomly ordered sets of digits, such as "5 8 2 1," the animals picked out the larger digit with a much higher success rate than chance alone would have predicted.

I cannot leave this topic without mentioning the curious difficulties that Sheba met when she had to pick the *smaller* of two numbers. The experimental situation seemed quite simple: The animal was shown two sets of food, and when it pointed to one, the experimenter gave it to another chimp while Sheba received the *other* food set. In this novel situation, it was in Sheba's interests to designate the smaller quantity, so that she would then receive the larger one. However, the chimpanzee never succeeded. She continued to point to the larger set, as if choosing the maximum amount of food was an irrepressible response. Sarah Boysen then thought of replacing the actual piles of food with the corresponding Arabic digits. Immediately, from the first trial, Sheba chose the smaller digit! Numerical symbols seemed to liberate Sheba from immediate material contingencies. They enabled her to act without being influenced by the parasitic impulse that otherwise compelled her to always pick out the larger amount of food.

The Limits of Animal Mathematics

How significant are such demonstrations of symbolic calculation in animals? Should they be viewed simply as circus acts extorted at the expense of an intensive training that turns animals into performing machines, but eventually tells us nothing about their normal abilities? Or are animals almost as gifted as humans in their ability to do mathematics? Without diminishing the importance of the above experiments, one is forced to admit that the mental manipulation of symbolic numerical labels in animals remains an exceptional finding. Although I have mentioned experiments with parrots, dolphins, and macaques, no cases of symbolic addition are known in any species other than the chimpanzee. Even their performance seems quite primitive when compared to that of a human child. It took Sheba several years of trial and error before she could master the digits 0 through 9. In the end, the chimpanzee still made frequent errors in using them, as did all the animals trained on number tasks. A young child, by contrast, spontaneously counts on its fingers, can often count up to 10 before the age of three, and rapidly moves on to multidigit numerals whose syntax is much more complex. The developing human brain seems to absorb language effortlessly—quite the opposite of animals, which always seem to need hundreds of repetitions of the same lesson before they retain anything.

What should we therefore remember about animal arithmetic? First, an undisputed and widespread ability to apprehend numerical quantities, to memorize, to compare, and even to add them approximately. Second, a considerably lesser ability, probably confined to a few species, for associating a repertoire of more or less abstract behaviors, such as pointing to an Arabic digit, to numerical representations. These behaviors may eventually serve as labels for numerical quantities—the "symbols." It is as if some animals could learn to grade the levels of the internal accumulator that they use to represent numbers. A lengthy training period enables them to memorize a list of behaviors: If the level of the accumulator is between *x* and *y*, then point to digit "2"; if it is between *y* and *z*, then point to digit "3"; and so on. This may just be a list of conditioned behaviors that is only remotely related to the extraordinary fluency that humans show when using the word "two" in contexts as different as "two apples," "two and two equal four," or "two dozen." While we may marvel at animals' ability to manipulate approximate representations of numerical quantities, teaching them a symbolic language seems to go against their natural proclivities. Indeed, the acquisition of symbols in animals never occurs in the wild.

From Animal to Human

Evolution is a conservative mechanism. When a useful organ emerges through random mutations, natural selection works to pass it on to the next generations. Indeed, the preservation of favorable traits is a major source of the organization of life. Therefore, if our closest cousins, the chimpanzees, possess some competence for arithmetic, and if species as different as rats, pigeons, and dolphins are not devoid of numerical abilities, it is likely that we *Homo sapiens* have received a similar heritage. Our brains, like the rat's, are likely to come equipped with an accumulator that enables us to perceive, memorize, and compare numerical magnitudes.

Many outstanding differences separate human cognitive abilities from those of other animals, including chimpanzees. For one thing, we have an uncanny ability to develop symbol systems, including a mathematical language. We are also endowed with a cerebral language organ that enables us to express our thoughts and to share them with other members of our species. Finally, our ability to devise intricate plans for actions, based on both a retrospective memory of past events and a prospective memory of future possibilities, seems to be unique in the animal kingdom. Does that mean, however, that in other respects, our cerebral hardware for number processing should be very different from that of other animals? The simple working hypothesis that I am defending throughout this book postulates that we are in fact endowed with a mental representation of quantities very similar to the one that can be found in rats, pigeons, or monkeys. Like them, we are able to rapidly enumerate collections of visual or auditory objects, to add them, and to compare their numerosities. I speculate that these abilities not only enable us to quickly work out the numerosity of sets, but also underlie our comprehension of symbolic numerals such as Arabic digits. In essence, the number sense that we inherit from our evolutionary history plays the role of a germ favoring the emergence of more advanced mathematical abilities.

In the next chapters, we will scrutinize human mathematical abilities, looking for vestiges of the animal mode of apprehending numbers. The first and perhaps the most dramatic cue that we will study is the remarkable competence of human infants in arithmetic, long before they first sit in a classroom—in fact, long before they can sit at all!

Chapter 2

Babies Who Count

The soul, as being immortal, and having been born again many times, and having seen all things that exist, whether in this world or in the world below, has knowledge of them all; and it is no wonder that she should be able to call to remembrance all that she ever knew about virtue, and about everything.

Plato, *Meno*

Do babies have any abstract knowledge of arithmetic at birth? The question seems preposterous. Intuition suggests that babies are virgin organisms initially devoid of any kind of competence other than the ability to learn. Yet if our working hypothesis is correct, the human brain is endowed with an innate mechanism for apprehending numerical quantities, one that is inherited from our evolutionary past and that guides the acquisition of mathematics. To influence the learning of number words, this protonumerical module must be in place before the period of exuberant language growth that some psychologists call the "lexical explosion," which occurs around a year and a half of age. In the first year of life, then, babies should already understand some fragments of arithmetic.

Baby Building: Piaget's Theory

Only over the last fifteen years has the subject of babies' numerical competence been examined empirically. Before this period, developmental psychology was dominated by constructivism, a view of human development that made the very notion of arithmetic in the first year of life sound inconceivable. According to the theory first set forth some fifty years ago by Jean Piaget, the founder of constructivism, logical and mathematical abilities are progressively constructed in the baby's mind by observing, internalizing, and abstracting regularities about the external world. At birth, the brain is a blank page devoid of any conceptual knowledge. Genes do not grant the organism any abstract ideas about the environment in which it will live. They merely instill simple perceptual and motor devices and a general learning mechanism that progressively takes advantage of the interactions of the subject with its environment to organize itself.

In the first year of life, according to constructivist theory, children are in a "sensorimotor" phase: they explore their environment through the five senses, and they learn to control it through motor actions. In this process, Piaget argues that children cannot fail to notice certain salient regularities. For instance, an object that disappears behind a screen always reappears when the screen is lowered; when two objects collide, they never interpenetrate; and so on. Guided by such discoveries, babies progressively construct a series of ever more refined and abstract mental representations of the world in which they are growing up. In this view, then, the development of abstract thought consists in climbing a series of steps in mental functioning, the Piagetian stages, that psychologists may identify and classify.

Piaget and his colleagues speculated a good deal about how the concept of number develops in young children. They believed that number, like any other abstract representation of the world, must be constructed in the course of sensorimotor interactions with the environment. The theory goes something like this: Children are born without any preconceived idea about arithmetic. It takes them years of attentive observation before they really understand what a number is. By dint of manipulating collections of objects, they eventually discover that number is the only property that does not vary when objects are moved around or when their appearance changes. Here is how Seymour Papert, in 1960, described this process:

> For the infant, objects do not even exist; an initial structuration is needed to organize experience into *things*. Let us stress that the baby does not *discover* the existence of objects like an explorer discovers a mountain, but rather like someone discovers music: he has heard it for years, but before then it was only noise to his ears. Having "acquired objects," the child still has a long way to go before reaching the stage of classes, seriations, inclusions and, eventually, number.

Piaget and his many collaborators had seemingly collected proof upon proof of young children's inability to understand arithmetic. For instance, if you hide a toy under a cloth, ten-month-old babies fail to reach for it—a finding that Piaget thought meant that babies believe that the toy ceases to exist when it is out of sight. Would this apparent lack of "object permanence," in Piagetian jargon, not imply that babies are fully ignorant of the world they live in? If they do not realize that objects continue to exist when they are out of sight, how could they ever know anything about the more abstract and evanescent properties of number?

Other observations by Piaget seemed to indicate that the number concept does not begin to be understood before the ages of four or five. Before then, children fail in what Piaget called the "number conservation" test. First, they are shown equally spaced rows of six glasses and six bottles. If they are now asked whether there are more glasses or more bottles, children will reply, "It's the same thing." They apparently rely on the one-to-one correspondence between objects in the two rows. The row of glasses is then spread so that it becomes longer that the row of bottles. Obviously, number is not affected by this manipulation. Yet when the earlier question is repeated, children now systematically respond that there are more glasses than bottles. They do not seem to realize that moving the objects around leaves their number unchanged. Psychologists would say that they do not "conserve number."

When children eventually pass the number conservation test, constructivists still do not grant them much conceptual understanding of arithmetic. Until they are seven or eight, it is still easy to entrap them with simple numerical tests. Show them, for instance, a bunch of eight flowers with six roses and two tulips, and ask them a silly question: Are there more roses or more flowers? Most of them will tell you that the roses are more numerous than the flowers! And Piaget readily concludes that prior to the age of reason, children lack knowledge of the most elementary bases of set theory, which many mathematicians believe to provide a foundation for arithmetic: They seemingly ignore that a subset cannot have more elements than the original set from which it was drawn.

Piaget's findings have had a considerable impact on our education system. His conclusions have instilled a pessimistic attitude and a wait-and-see policy among educators. The theory states that the regular climbing of Piagetian stages progresses according to an immutable process of growth. Before the age of six or seven, the child is not "ready" for arithmetic. Hence, precocious teaching of mathematics is a vain or even harmful enterprise. If it is taught early on, the number concept cannot but be distorted in kids' heads. It will have to be learned by rote, without any genuine understanding. Failing to grasp what arithmetic is about, children will develop a strong feeling of anxiety about mathematics.

According to Piagetian theory, it is best to start by teaching logic and the ordering of sets, because these notions are a prerequisite to the acquisition of the concept of number. This is the main reason why, even today, children in most preschools spend much of their day piling up cubes of decreasing sizes, long before they learn to count.

Is such pessimism reasonable? We have seen that rats and pigeons readily recognize a certain number of objects, even as their spatial configuration varies. We know already that a chimp will spontaneously choose the larger of two numerical quantities. Is it conceivable that human children before the age of four or five lag so far behind other animals in arithmetic?

Piaget's Errors

We now know that this aspect of Piaget's constructivism was wrong. Obviously, young children have much to learn about arithmetic, and obviously their conceptual understanding of numbers deepens with age and education—but they are not devoid of genuine mental representations of numbers, even at birth! One merely has to test them using research methods tailored to their young age. Unfortunately the tests that Piaget favored do not enable children to show what they are really capable of. Their major defect lies in their reliance on an open dialog between experimenters and their young subjects. Do children really understand all the questions that they are being asked? Most important, do they interpret these questions as adults would? There are several reasons to think not. When children are placed in situations analogous to those used with animals and when their minds are probed without words, their numerical abilities turn out to be nothing less than considerable.

Take, for instance, the classical Piagetian test of number conservation. As early as 1967, in the prestigious scientific journal *Science*, Jacques Mehler and Tom Bever, then at the department of psychology at MIT, demonstrated that the results of this test changed radically according to context and to the children's level of motivation. They showed the same children, two to four years old, two series of trials. In one—similar to the classical conservation situation—the experimenter set up two rows of marbles. One row was short but consisted of six marbles, and the other, although longer, had only four marbles (Figure 2.1). When the children were asked which row had more marbles, most three- and four-year-olds got it wrong and selected the longer but less numerous row. This recalls Piaget's classical nonconservation error.

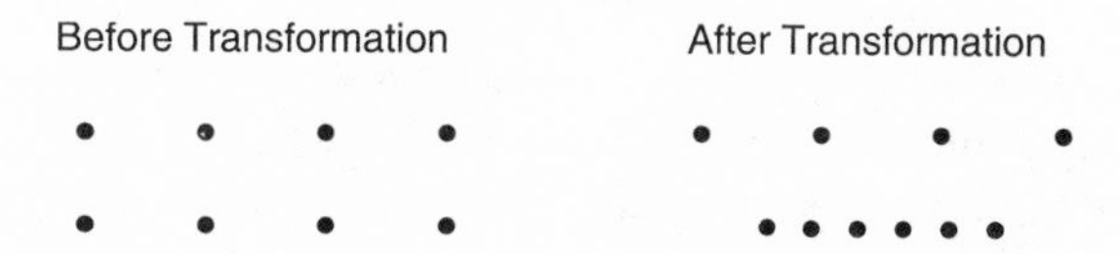

Figure 2.1. When two rows of items are in perfect one-to-one correspondence (left panel), a three- or four-year-old child states that they are equal. If one now transforms the bottom row both by shortening it and by adding two items (right panel), the child declares that the top *row has more items. This is the classical error first discovered by Piaget: The child responds on the basis of row length rather than number. Yet when the rows are made up of M&Ms, Mehler and Bever (1967) proved that children spontaneously choose the* bottom *row. Hence, the Piagetian error is not imputable to children's incompetence in arithmetic, but merely to the disconcerting conditions of number conservation tests. (After Mehler and Bever, 1967.)*

In the second series of trials, however, Mehler and Bever's ruse consisted in replacing marbles with palatable treats (M&Ms). Instead of being asked complicated questions, the children were allowed to pick up one of the two rows and consume it right away. This procedure had the advantage of sidestepping language comprehension difficulties while increasing the children's motivation to choose the row with the most treats. Indeed, when candy was used, a majority of children selected the larger of the two numbers, even when the length of the rows conflicted with number. This provided a striking demonstration that their numerical competence is no more negligible than their appetite for sweets!

That three- and four-year-old children select the more numerous row of candy is perhaps not very surprising, even though it conflicts directly with Piaget's theory. But there is more. In Mehler and Bever's experiment, the youngest children, who were about two years old, succeeded perfectly in the test, both with marbles and with M&Ms. Only the older children failed to conserve the number of marbles. Hence, performance on number conservation tests appears to drop temporarily between two and three years of age. But the cognitive abilities of three- and four-year-olds are certainly not less well-developed than those of two-years-olds. Hence, Piagetian tests cannot measure children's true numerical competence. For some reason, these tests seem to confuse older children to such an extent that they become unable to perform nearly as well as their younger brothers and sisters.

I believe that what happens is this: Three- and four-year-olds interpret the experimenter's questions quite differently from adults. The wording of the questions and the context in which they are posed mislead children into believing that they are asked to judge the length of the rows rather their numerosity. Remember that, in Piaget's seminal experiment, the experimenter asks the very same question twice: "Is it the same thing, or does one row have more marbles?" He first raises this question when the two rows are in perfect one-to-one correspondence, and then again after their length has been modified.

What might children think of these two successive questions? Let us suppose for a moment that the numerical equality of the two rows is obvious to them. They must find it quite strange that a grown-up would repeat the same trivial question twice. Indeed, it constitutes a violation of ordinary rules of conversation to ask a question whose answer is already known by both speakers. Faced with this internal conflict, perhaps children figure out that the second question, although it is superficially identical to the first, does not have the same meaning. Perhaps something like the following reasoning goes on in their heads:

> If these grown-ups ask me the same question twice, it must be because they are expecting a different answer. Yet the only thing that changed relative to the previous situation is the length of one of the rows. Hence, the new question must bear on the length of the rows, even though it seems to bear on their number. I guess I'd better answer on the basis of row length rather than on the basis of number.

This line of reasoning, although quite refined, is well within the reach of three- and four-year-olds. In fact, unconscious inferences of this type underlie the interpretation of a great many sentences, including those that a very young child may produce or comprehend. We all routinely perform hundreds of inferences of this sort. Understanding a sentence consists in going beyond its literal meaning and retrieving the actual meaning initially intended by the speaker. In many circumstances, the actual meaning can be the direct opposite of the literal sense. We speak of a good movie as being "not too bad, isn't it?" And when we ask, "Could you pass the salt?" we are certainly not satisfied when the answer is a mere "yes"! Such examples demonstrate that we constantly reinterpret the sentences that we hear by performing complex unconscious inferences concerning the other speaker's intentions. There is no reason to think that young children are not doing the same when they converse with an adult during these tests. In fact, this hypothesis seems all the more plausible since it is precisely around three or four years of age—the point at which Mehler and Bever find that children begin not to conserve number—that the ability to reason about the intentions, beliefs, and knowledge of other people, which psychologists call a "theory of mind," arises in young children.

Two developmental psychologists from the University of Edinburgh, James McGarrigle and Margaret Donaldson, directly tested the hypothesis that children's failure to "conserve number" on Piagetian tests is linked to their misunderstanding of the experimenter's intentions. In their experiment, half of the trials were of the classical type, where the experimenter modified the length of one row and then asked, "Which has more?" In the other half of the trials, however, the length transformation was performed fortuitously by a teddy bear. While the experimenter was conveniently looking elsewhere, a teddy bear lengthened one of the two rows. The experimenter then turned and exclaimed, "Oh no! The silly teddy bear has again mixed up everything." Only then did the researcher again ask the question "Which has more?" The underlying idea was that, in this situation, this query seemed sincere and could be interpreted in a literal sense. Since the bear had messed up the two rows, the adult did not know anymore how many objects there were and hence was asking the child. In this situation the vast majority of children responded correctly on the basis of number, without being influenced by row length. The same children, however, failed by systematically responding on the basis of length when the transformation was performed intentionally by the experimenter. This proves two points: First, even a young child is capable of interpreting the same exact question in two quite different ways depending on context. Second, Piaget notwithstanding, when the question is asked in a context that makes sense, young children get the answer right—they can conserve number!

I would not want to leave this discussion on a misunderstanding. I certainly do not consider the children's failure on Piagetian conservation tasks to be a trivial matter. On the contrary, this is an active domain of research that still attracts many researchers throughout the world. After hundreds of experiments, it is still unclear exactly why children are so easily deceived by fallacious cues such as row length when they have to judge number. Some scientists think that failure on Piagetian tasks reflects the continuing maturation of the prefrontal cortex, a region of the brain that enables us to select a strategy and to hold firm to it despite distraction. If this theory turns out to be correct, Piagetian tests could take on a new meaning as a behavioral marker of children's ability to resist distraction. However, developing such ideas would be the matter of another book. My purpose, here, is more modest. My sole objective is to convince you that we now know what Piagetian tests are *not* about. Contrary to what their inventor thought, these are not good tests of when a child begins to understand the concept of number.

Younger and Younger

The experiments that I have described so far challenge the Piagetian time scale for numerical development by suggesting that children "conserve number" at a much earlier age than was once thought possible. Yet do they refute the whole of constructivism? Not really. Piaget's theory is much more subtle than I can possibly describe in a few paragraphs, and it allows several ways in which he might have accommodated the above results.

He might have argued, for instance, that by removing some of the conflicting cues from his original number conservation test, the modified experiments made the children's task too simple. Piaget was well aware that his number conservation test misled children—in fact it was *purposely* designed so that row length conflicted with number. In his view, children really mastered the conceptual underpinnings of arithmetic only when they could predict which row had the most items on a purely logical basis, by reflecting on the logical consequences of the operations that had occurred, and without letting themselves be distracted by irrelevant changes in row length or in the way the experimenter phrased the questions. Resistance to misleading cues, it seems, was part and parcel of Piaget's definition of what it meant to have a conceptual understanding of number.

Piaget might also have argued that choosing the largest number of candy does not require a *conceptual* understanding of number, but only a sensorimotor coordination that allows the child to recognize the greater pile and orient to it. Throughout his work, Piaget ceaselessly stressed young children's sensorimotor intelligence, so he might well have happily accepted that children discovered the "choose-the-larger" strategy at an early age. He would have insisted, however, that this strategy was used without any understanding of its logical basis; only later, he claimed, would children reflect on their sensorimotor abilities and arrive at a more abstract construal of number. Typical of this attitude is Piaget's reaction when he heard about Otto Koehler's work on perception of numerosity in birds and squirrels—he accepted that animals could acquire "sensorimotor numbers," but not a conceptual understanding of arithmetic.

Before the 1980s, the experiments that challenged Piaget's theory did not really address his central hypothesis that babies were devoid of a genuine concept of number. After all, the youngest children who sat for Mehler and Bever's marble test were already two years old. This still left a long time for learning to have taken place. In this context, scientific studies of infants suddenly became of paramount theoretical importance. Could it be shown that even babies under one year of age already mastered some aspects of the number concept, before they had had any chance of abstracting them from interactions with the environment? The answer is yes. In the 1980s, numerical abilities were observed in six-month-old infants and even in newborns.

Obviously, in order to reveal numerical competence at such an early age, verbal questioning will not do. Scientists have therefore relied on babies' appeal for novelty. Any parent knows that when a baby sees the same toy over and over again, it eventually loses interest in it. At this point, introducing a new toy can revive its interest. This elementary observation—which is obviously in need of being replicated in the laboratory and in a tightly controlled situation—proves that the child has noted the difference between the first and the second toy. This technique can be extended to ask babies all sorts of questions. It is in this way that researchers have been able to demonstrate that, very early in life, babies and even newborns can perceive differences in color, shape, size, and, more to the point, number.

The first experiment to establish that babies recognize small numbers took place in 1980 in Prentice Starkey's laboratory at the University of Pennsylvania. A total of 72 babies, aged between 16 and 30 weeks, were tested. Each baby, seated on its mother's lap, faced a screen on which slides were projected (Figure 2.2). A video camera focusing on the babies' eyes filmed its gaze, enabling an associate, who was blind to the exact conditions of the experiment, to measure exactly how long the baby spent looking at each slide. When the baby started looking elsewhere, a new slide appeared on the screen. Initially, the content of the slides was essentially the same: two large black dots, more or less spread out horizontally from trial to trial. In the course of trials, the baby started to look more and more briefly at this repetitive stimulus. The slides were then changed without warning to new slides containing three black dots. Immediately, the baby started to fixate longer at these unexpected images. The fixation time, which was 1.9 sec-

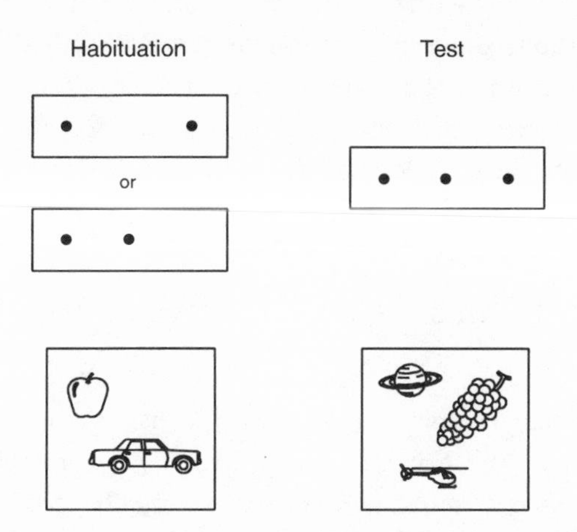

Figure 2.2. *To prove that infants discriminate the numerosities 2 and 3, they are first repeatedly shown collections with a fixed number of items, say two (left). Following this habituation phase, infants look longer at collections of three items (right) than at collections of two items. Because object location, size, and identity vary, only a sensitivity to numerosity can explain infants' renewed attention. (Top, stimuli used by Starkey and Cooper 1980; bottom, stimuli similar to those used by Strauss and Curtis 1981.)*

onds just prior to the switch, jumped to 2.5 seconds on the very first new slide. Hence, the baby detected the switch from two dots to three dots. Other children, tested in the same manner, detected the switch from three to two dots. Initially, these experiments were performed with six- or seven-month-olds, but a few years later Sue Ellen Antell and Daniel Keating, from the University of Maryland in Baltimore County, demonstrated with a similar technique that even newborns could discriminate numbers 2 and 3 a few days after birth.

How can one make sure that it is really the change in number that is noticed by the babies, rather than any other physical modification of the stimulus? In their initial experiments, Starkey and Cooper had aligned the dots so that the global figure that they formed provided no cue to number (in other arrangements, number is often confounded with shape because two dots form a line and three dots a triangle). They also varied the spacing between the dots so that neither their density, nor the total length of the line, would suffice to discriminate 2 from 3. Later, Mark Strauss and Lynne Curtis, at the University of Pittsburgh, introduced an even better control. They simply used color photographs of common objects of all kinds. The objects were small or large, aligned or not, and were photographed from near or far. Only their number remained constant: There were two objects in one half of the experiment, and three in the other. Not the least affected by such variability in all the possible physical parameters, the babies continued to notice the change in number. More recently, the experiment has even been replicated by Eric Van Loosbroek and Ad Smitsman, two psychologists from the Catholic University of Nijmegen in the Netherlands, with moving displays—random geometrical shapes that occasionally hide one another in the course of their trajectory. In the first few months of life, babies appear to notice the constancy of objects in a moving environment and extract their numerosity.

Babies' Power of Abstraction

It remains to be seen whether this precocious sensitivity to numerosity merely reflects the power of the babies' visual system, or whether it betrays a more abstract representation of number. With very young children, we have to ask the very same questions as those raised with rats and chimpanzees. Are they able to extract the number of tones in an auditory sequence, for instance? Most important, do they know that the same abstract concept "3" applies to three sounds and to three visual objects? Finally, can they mentally combine their numerical representations to perform elementary calculations such as 1+1=2?

To answer the first question, scientists simply moved the original experiments on the visual recognition of number to the auditory modality. They bored babies by repeating sequences of three sounds over and over again, and then ascertained whether a later novel sequence of two sounds was able to renew their interest. One of these experiments is especially instructive because it suggests that, as early as four days of age, a baby can decompose speech sounds into smaller units—syllables—that it can then enumerate. But at such a young age, it is easier to use sucking rhythm rather than gaze orientation as an experimental tool. So Ranka Bijeljac-Babic and her colleagues at the Laboratory for Cognitive Science and Psycholinguistics in Paris have babies suck on a nipple connected to a pressure transducer and a computer. Whenever the baby sucks, the computer notices it and immediately delivers a nonsense word such as "bakifoo" or "pilofa" through a loudspeaker. All the words share the same number of syllables—three, for instance. When a baby is first placed in this peculiar situation where sucking yields sound, it shows an increased interest which is translated into an elevated sucking rate. After a few minutes, however, sucking drops. As soon as the computer detects this drop, it switches to delivering words with only two syllables. The baby's reaction: It immediately goes back to sucking vigorously in order to listen to the new word structure. To ensure that this reaction is related to the number of syllables rather than to the mere presence of novel words, with some babies novel words are introduced while the number of syllables is left unchanged. In this control group, no reaction is perceptible. Since the duration of words and the rate of speech are highly variable, the number of syllables is indeed the only parameter that can enable babies to differentiate the first list of words from the second.

Very young children therefore pay equal attention to the number of sounds as to the number of objects in their environment. We also know, thanks to a recent experiment by Karen Wynn, that at six months of age they will discriminate numbers of actions, such as a puppet making two jumps versus three jumps. Yet are they aware of the "correspondence" between sound and sight, to paraphrase the French poet Baudelaire? Do they anticipate that three strokes of lightning should predict an equal number of thunderclaps? In brief, do they access an abstract representation of number, independent of the visual or auditory modality that mediates it? Thanks to remarkably clever experiments designed by American psychologists Prentice Starkey, Elizabeth Spelke, and Rochel Gelman, we can now give a positive answer to this question. I rank their work highly on my personal pantheon of experimental psychology because only fifteen years ago it would have seemed virtually impossible to ask such a complex question about a baby's mind.

In this multimedia experiment, a six-, seven-, or eight-month-old baby is seated in front of two slide projectors. On the right, the slide shows two common objects, randomly arranged. On the left, a similar slide shows three objects. Simultaneously, the baby hears a sequence of drum beats played by a central loudspeaker placed between the two screens. Finally, as usual the baby is watched by a hidden video camera that enables experimenters to measure how much time the baby spends looking at each slide.

Initially, the baby is attentive and explores the images visually. Obviously those with three objects are more complex than those with only two, so the baby dedicates a little more time and attention to them. After a few trials, however, this bias fades, and a fascinating result emerges: The baby looks longer at the slide whose numerosity matches the sequence of sounds that it is hearing. It consistently looks longer at three objects when hearing three drumbeats, but now prefers to watch *two* objects when hearing two drumbeats.

It therefore seems likely that the baby can identify the number of sounds—even though it varies from trial to trial—and is capable of comparing it to the number of objects before its eyes. If the two numbers are mismatched, the baby decides not to delve any longer into this slide, but rather to take a peek at the other one. The very fact that a child only a few months of age applies a strategy as sophisticated as this implies that its numerical representation is not tied to a low level of visual or auditory perception. The simplest explanation is that the child really perceives numbers rather than auditory patterns or geometrical configurations of objects. The very same representation of number "3" seems to fire in its brain whether it sees three objects or hears three sounds. This internal, abstract and amodal representation enables the child to notice the correspondence between the number of objects on one slide and the number of sounds that are simultaneously heard. Remember that animals behave in a very similar way: They too seem to possess neurons that respond equally well to three sounds or three light flashes. Babies' behavior may well reflect an abstract module for number perception, implanted by evolution ages ago, deep within the animal and human brains.

How Much Is 1 plus 1?

Let us momentarily pursue the comparison between the behavior of babies and that of other animal species. We have seen in the preceding chapter that a chimpanzee can compute the approximate total of a simple addition such as two oranges plus three oranges. Might this also be true of young infants? At first sight, this seems a rather daring hypothesis. We are more inclined to think that the acquisition of mathematics starts in the preschool years. It was not until the 1990s that a question as iconoclastic as the existence of calculation abilities in the first year of life received an empirical evaluation. By then, the scientific community had been sufficiently prepared by the many experiments on numerical perception, both in infants and in animals, for an experiment of this type to be attempted and for its results to receive attention.

In 1992, Karen Wynn's famous article on addition and subtraction by four- and five-month-old infants appeared in the journal *Nature*. The young American scientist had employed a simple yet ingenious design that relied on infants' ability to detect physically impossible events. Several earlier experiments had shown that, in their first year of life, infants express strong puzzlement when they witness "magical" events that violate the fundamental laws of physics. For instance, if they see an object remain mysteriously suspended in midair after losing its support, babies watch this scene with incredulous attention. Likewise, they express surprise when a scene suggests that two physical objects occupy the same location in space. Finally, if one hides an object behind a screen, babies find it astounding not to see it again when the screen later drops. In passing, note that this observation proves that, as early as five months and contrary to Piaget's theory, "out of sight" is not "out of mind." We now know that the failure of children under one year in Piaget's object permanence task is linked to the immaturity of their prefrontal cortex, which controls their reaching movements. The fact that they can't reach properly toward a hidden object does not imply that they believe it to be gone.

In all such situations, infants' surprise is demonstrated by a significant increase in the amount of time they spend examining the scene, relative to a control situation in which the laws of physics have not been violated. Karen Wynn's knack resides in adapting this idea to probe infants' number sense. She showed them events that could be interpreted as numerical transformations—for instance, one object plus another object—and tested whether infants expect the precise numerical outcome of two objects.

Upon arrival in the laboratory, the five-month-old participants discovered a little puppet theater with a rotating screen up front (Figure 2.3). The hand of the experimenter came out on one side, holding a toy Mickey Mouse, which it placed on stage. Then the screen came up, masking the location of the toy. The hand appeared on the scene a second time with a second Mickey Mouse, deposited it behind the screen and left empty. The entire sequence of events stood for a concrete depiction of the addition 1+1: Initially, there was only one toy behind the screen, and then a second one was added. Children never saw the two toys together, but only one after the other. Would they have inferred, nevertheless, that there should be two Mickeys behind the screen?

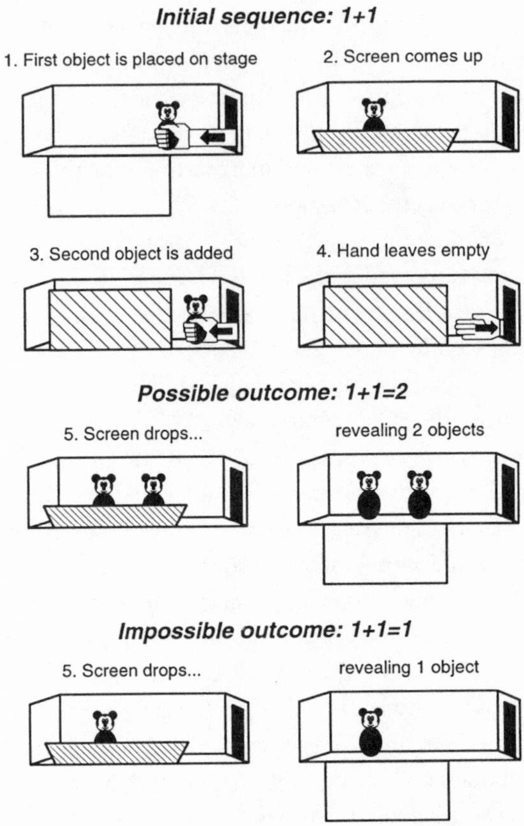

Initial sequence: 1+1

1. First object is placed on stage
2. Screen comes up
3. Second object is added
4. Hand leaves empty

Possible outcome: 1+1=2

5. Screen drops... revealing 2 objects

Impossible outcome: 1+1=1

5. Screen drops... revealing 1 object

Figure 2.3. Karen Wynn's experiment shows that 4½-month-olds expect 1 plus 1 to make 2. First, a toy is hidden behind a screen. Then a second identical toy is added. Finally, the screen drops, sometimes revealing the two toys, and sometimes only one (the other toy having been surreptitiously taken away). Infants look systematically longer at the impossible event "1+1=1" than at the possible one "1+1=2," suggesting that they were expecting two objects. (Adapted from Wynn, 1992.)

To figure this out, the screen was lowered, revealing an unexpected result: Only one Mickey could be seen! Unbeknownst to the subjects, one of the two toys had been removed through a hidden trap door. In order to estimate the infants' degree of surprise, the time that they spent fixating this impossible situation "1+1=1" was measured and compared to the fixation time for the expected outcome of two objects ("1+1=2"). On average, infants looked one second longer at the false addition 1+1=1 than at the possible event 1+1=2. One might still object that the kids were not really computing additions, but were simply looking longer at a single object than at two identical ones. However, this explanation is not be tenable, because the results were reversed in a second group of babies who were presented with the operation 2–1 instead of 1+1. In this group, the babies were now surprised to discover two objects behind the screen (2–1=2), and they examined this situation as much as three seconds longer than the possible event 2–1=1.

As Wynn herself observes, if one wants to play the devil's advocate, these results still need not imply that babies can perform exact computations. They may just know that the numerosity of a set changes when objects are added or removed. Hence, they might figure out that 1+1 cannot possibly equal 1, nor 2–1 equal 2, without necessarily knowing the exact result for these operations. Yet even this contrived explanation does not stand up to empirical testing. One merely has to replicate the addition situation 1+1 with outcomes of either two or three objects. Karen Wynn ran this replication and observed that, again, five-month-old babies looked longer at the impossible outcome of three objects than at the possible outcome of two objects. The demonstration is irrefutable: Babies know that 1+1 makes neither 1 nor 3, but exactly 2.

This knowledge puts infants on a par with the rats we looked at, or with Sheba, the chimp prodigy whose computing abilities were described in the previous chapter. In fact, the exact design of Karen Wynn's experiment has now been replicated by Harvard psychologist Mark Hauser with rhesus monkeys in the wild. When a monkey, intrigued by Hauser's presence, volunteered to look at him, Hauser successively hid two eggplants in a box. Then, in some trials only, he sneaked one off before opening the box, while a colleague filmed the animal to measure its degree of surprise. The results of this wild scene were important and fascinating. The monkeys reacted even more strongly than babies: On the "magic" trials in which one of the expected eggplants was missing, they spent considerable time scrutinizing the box. Obviously, human infants are at least as gifted as their animal cousins in arithmetic, confirming that elementary numerical computations can be performed by organisms devoid of language.

Still, Karen Wynn's experiments give no clue as to how abstract infants' knowledge really is. Infants may keep a vivid and realistic image of the objects hidden behind the screen—a kind of mental photograph sufficiently precise for them to immediately notice any missing or supernumerary objects. Alternatively, they may only keep a memory of the number of objects added to or subtracted from behind the screen, without caring about their location and identity. To find out, one may prevent children from building a precise mental model of the objects' location and identity, and see whether they can still anticipate their number. This idea has served as the basis for an experiment recently conducted by Etienne Koechlin in our laboratory in Paris. The design is quite similar to Wynn's studies, except that objects are now placed on a slowly rotating turntable that keeps them in constant motion even when they are hidden behind the screen. It is therefore impossible to predict where they will be when the screen drops. Babies cannot conjure up a precise mental image of the predicted scene; all they can construct is an abstract representation of two rotating objects with unpredictable locations.

The results, amazingly, show that four-and-a-half-month-old infants are not in the least confused by object motion. They still find the impossible events 1+1=1 and 2−1=2 surprising. Hence their behavior does not depend crucially on the expectation of precise object locations. They do not expect to find a precise configuration of objects behind the screen, but merely two objects—no more, no less. A psychologist at the Georgia Institute of Technology, Tony Simon, and his colleagues have even shown that infants do not attend to the exact identity of the objects behind the screen when computing their number. Unlike older children, four- and five-month-olds are not surprised much by changes in object appearance in the course of arithmetical operations. If two Mickey Mouse toys are placed behind the screen, they are not shocked to discover two red balls instead of the original toys when the screen drops. Yet their attention is highly aroused if only one ball is to be seen. Mickey Mouse turning into a ball, or the toad changing into a prince, is an acceptable transformation as far as the baby's number processing system is concerned. As long as no object vanishes or is created *de novo*, the operation is judged to be numerically correct and yields no surprise reaction in babies. In contrast, the disappearance of an object or its inexplicable replication, as in the miracle of the loaves and fishes, seems miraculous because it violates our deepest numerical expectations. Not only is keeping track of a small number of objects child's play, but the child's number sense is sufficiently sophisticated to avoid being deceived by object motion or by sudden changes in object identity.

The Limits of Infant Arithmetic

I hope that these experiments have convinced you that young children have natural talent for numbers. This does not mean, however, that you should enroll your youngest toddler in evening math classes. Neither do I recommend consulting a child neurologist if your kids make astronomical mistakes in elementary additions. Shame on me if my rebuttal of Piaget has served as a pretext for the charlatans who claim they can arouse intelligence in the first year of life by presenting infants with additions written in Arabic digits or even with Japanese characters, which they are of course totally unable to understand. While young children's numerical abilities are real, they are strictly limited to the most elementary of arithmetic.

In the first place, their abilities for exact calculation do not seem to extend beyond the numbers 1, 2, 3, and perhaps 4. Whenever experiments involve sets of two or three objects, infants are found to discriminate them. However, only occasionally are they shown to differentiate three versus four. And never can a group of babies under one year of age distinguish four dots from five or even from six. Apparently, babies only have an accurate knowledge of the first few numbers. Their competence, in this domain, may well be inferior to that of adult chimpanzees, whose performance remains above chance even when they have to choose between six versus seven pieces of chocolate.

Let us not jump too quickly to the conclusion that number 4 marks the confines of the baby's arithmetic universe. The experiments available to date have concentrated on the exact representation of small integers in the baby's mind. Babies, however—like rats, pigeons, or monkeys—most likely possess only an approximate and continuous mental representation of numbers. This representation probably obeys the distance and size effects found in rats and in chimpanzees. We should therefore expect babies to be unable, beyond some limit, to discriminate a number n from its successor $n+1$. This is indeed what is observed beyond number 4. However, we should also expect them to recognize numbers beyond this limit, provided that they are contrasted with even more distant numbers. Thus, babies may not know whether 2+2 is 3, 4, or 5, yet they may still be surprised if they see a scene suggesting that 2+2 is 8. To my knowledge, this prediction has not yet been put to a test. If proved correct, it would considerably extend the numerical knowledge attributed to very young children.

Baby arithmetic has a second major limitation. In situations where an adult would automatically infer the presence of several objects, babies do not necessarily draw the same conclusion. Let me explain. Suppose that you alternatively see a small red truck and then a green ball popping out from behind a screen. You would immediately conclude that at least two objects are hiding there, and you would be much puzzled to discover only one object, say the green ball, when the screen is removed. Young children react differently. Whether one or two objects are visible when the screen drops, ten-month-old infants do not show any sign of surprise. Apparently, babies do not consider the fact that quite different shapes and colors alternatively come out from behind the screen as a sufficient clue to the presence of several objects. Babies fail even when the experiment is performed with highly familiar objects such as the subjects' own bottle or their favorite doll. Only at twelve months of age do they start to expect two objects. Even then, the experiment works only with objects of different shapes. If only color or size varies, even a twelve-month-old thinks that seeing a large ball popping out of one side of a screen and a small one on the other side is not sufficient to infer the presence of two different objects behind the screen.

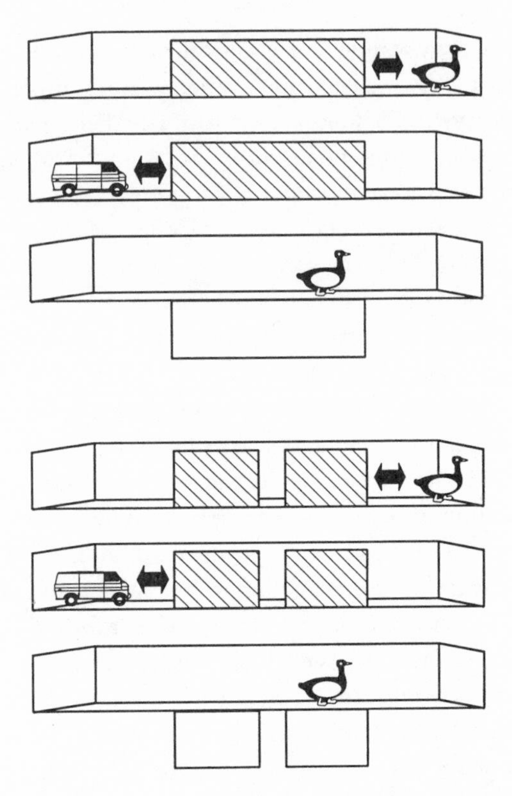

Figure 2.4. *Infants' numerical expectations are based on object trajectory, not on object identity. In the top situation, a duck and a truck alternately appear at the right and left of a screen. Despite the change in object identity, infants show no surprise when the screen drops and reveals a single object. In the bottom situation, a window is cut in the screen, making it physically impossible for an object to move from right to left without appearing in this window for a short while. In this situation, infants expect two objects and are surprised if only one is found when the screen drops. (Adapted from Xu and Carey, 1996.)*

The only clue that babies seem to find conclusive is the trajectory followed by objects (Figure 2.4). Thus, when the same experiment is repeated with not just one, but *two* screens separated by a void, if an object alternatively pops out from the right screen and from the left screen, babies infer the presence of two objects, one behind each screen. They know that it would be impossible for a single object to move from one screen to the other without appearing even for a short moment in the space separating them. If, however, an object does appear in this space at the appropriate time, then the babies' preference switches, and they again expect only one object. And conversely, if there is only one screen but the babies are shown the two objects together on the stage for only two seconds at the beginning of the experiment, then they expect to find two objects at the end.

Information about the spatial trajectories of objects thus provides a crucial cue to numerosity perception. Note that this conclusion does not contradict in any way the results of the turntable experiment I described above, which showed that babies did not care whether the objects behind the screen moved or stood still. In fact, there is every reason to believe that in that experiment too, trajectory information is crucial. In the "1+1=2" condition, for instance, just after a first Mickey Mouse toy has been placed on the turntable behind the screen, an identical toy appears in the experimenter's hand to the right of the screen. It is physically impossible for it to be the same toy as before, for this toy could not possibly leave from behind the screen without being seen. Hence, infants conclude that there is a second Mickey superficially identical to the first—and therefore expect a total of two objects. It does not matter if the toys are subsequently moved around until their locations are unpredictable. Once the abstract representation of "2" has been activated, it can resist this type of modification. Spatial information about the location of discrete objects in space and time is critical to set up the representation of number in the baby's brain; but it is not needed once this representation has been activated.

In summary, babies' numerical inferences seem to be completely determined by the spatiotemporal trajectory of objects. If the motion that they see could not possibly be caused by a single object without violating the laws of physics, they draw the inference that there are at least two objects. Otherwise, they stick to the default hypothesis that there is only one object, even if that implies that the object is constantly changing in shape, size, and color. Thus, the baby's numerical module is both hypersensitive to information about object trajectory, location, and occlusion, and completely blind to changes in shape or color. Never mind the identity of the object; only location and trajectory matter.

Only a rather foolish detective neglects half of the available cues. Because we have been accustomed to a much higher standard of performance in babies, however, we have to ask whether this strategy is not more clever than it appears. Is the baby's line of reasoning deficient, or does it attest, on the contrary, a wisdom worthy of a Sherlock Holmes? After all, everyone knows that a criminal can disguise himself to be a number of different people. Such is also the case with many common objects whose aspect varies. The profiles and faces of people, for instance, are very dissimilar visual objects, yet babies have to learn that they are merely different views of the same persons. How could a child know beforehand that a truck cannot turn itself into a ball, while a tiny piece of red rubber readily transforms itself into a big pink balloon when someone blows in it? This kind of anecdotal information cannot be known in advance. It has to be learned piece by piece on each encounter with a new object. Yet in order to learn something, one must not be too prejudiced. This might explain why babies default on the hypothesis that only one object is out there. As good logicians, they maintain this hypothesis until there is clear proof to the contrary, even if they witness curious transformations in object shape and color.

From an evolutionary viewpoint, it is rather remarkable that nature founded the bases of arithmetic on the most fundamental laws of physics. At least three laws are exploited by the human "number sense." First, an object cannot simultaneously occupy several separate locations. Second, two objects cannot occupy the same location. Finally, a physical object cannot disappear abruptly, nor can it suddenly surface at a previously empty location: Its trajectory has to be continuous. We owe child psychologists Elizabeth Spelke and Renée Baillargeon the discovery that even very young babies understand these laws. Indeed, in our physical environment, they admit very few exceptions, the most prominent being caused by shadows, reflections, and transparencies. (Perhaps this may explain the fascination and the confusion that these "objects" exert on young children.) These principles therefore provide a firm foundation for the small amount of number theory that the animal and human brains seems to be endowed with. The infant brain relies exclusively on them to predict how many distinct objects are present. It stubbornly refuses to exploit other cues to number that may be accidental, such as the visual appearance of objects. This attests to the antiquity of babies' "number sense," for only evolution, with its millions of years of trial and error, could possibly sort out the fundamental and the anecdotal properties of physical objects.

Indeed, the tight link between discrete physical objects and numerical information endures up to a much older age, where it eventually has a negative impact on some aspects of mathematical development. If you know a three- or four-year-old child, try the following experiment. Show him the picture on Figure 2.5 and ask him how many forks he can see. You will be surprised to discover that he reaches an erroneous total, because he counts every single piece of a fork as one unit. He counts the broken fork twice and announces a total of six. It is extremely

Figure 2.5. Three- to four-year-olds believe that this set comprises six forks. They cannot avoid counting each discrete physical object as one unit. (Adapted from Shipley and Shepperson 1990.)

difficult to explain to him that the two separate pieces should be counted as one unit. Likewise, show him two red apples and three yellow bananas, and ask him how many different colors there are, or how many different kinds of fruit he can see. Obviously, the correct response is two. Yet up to a relatively advanced age, children cannot help counting every single object as one unit and therefore reach the erroneous total of five. The maxim "Number is a property of sets of discrete physical objects" is deeply embedded in their brains.

Nature, Nurture, and Number

Throughout this chapter, I have spoken of babies as though they were inert organisms with rigid performances. When discussing experiments with young children, we easily forget that age groups can vary from a few days up to ten or twelve months of age. In fact, the first year of life is when the baby's brain possesses maximal plasticity. During this period, babies absorb an impressive amount of new knowledge day after day and can therefore hardly be considered as a static system whose performance is stable. Right after birth, they learn to recognize their mother's voice and face; they begin to process the language spoken in their surroundings; they discover how to command their body movements; and the list could go on forever. We have no reason to believe that numerical development escapes this general outburst of learning and discovery.

To do justice to the fluidity of babies' intelligence, the numerical abilities that I have described in this chapter should be situated within a dynamic framework —a perilous exercise given that we still know so little about the logic with which the representation of number evolves in the first year of life. But at least we can try to sketch a tentative scenario of the order and the way in which these abilities mature with the passing months.

Let us start with birth, an age at which number discrimination abilities have already been amply demonstrated. Newborns readily distinguish two objects from three and perhaps even three from four, while their ears notice the difference between two and three sounds. Hence, their brain apparently comes equipped with numerical detectors that are probably laid down before birth. The plan required to wire up these detectors probably belongs to our genetic endowment. Indeed, it is hard to see how children could draw from the environment sufficient information to learn the numbers 1, 2, and 3 at such an early age. Even supposing that learning is possible before birth or in the first few hours of life—during which visual stimulation is often close to nil—the problem remains, because it seems impossible for an organism that ignores all about numbers to learn to recognize them. It is as if one asked a black-and-white TV to learn about colors! More likely, a brain module specialized for identifying numbers is laid down through the spontaneous maturation of cerebral neuronal networks, under direct genetic control and with minimal guidance from the environment. Since the human genetic code is inherited from millions of years of evolution, we probably share this innate protonumerical system with many other animal species—a conclusion whose plausibility we have judged in the preceding chapter.

Though the newborn may be equipped with visual and auditory numerosity detectors, no experiment to date proves that these two input modalities communicate and share their numerical cues right from birth. At present, only in six- to eight-month-old babies has the connection between two sounds and two images or three sounds and three images been demonstrated. While waiting for conclusive experiments with younger children, it remains possible to maintain that learning, rather than brain maturation, is responsible for the baby's knowledge of numerical correspondence between sensory modalities. By dint of hearing single objects emit only one sound, pairs of objects emit two sounds and so on, the baby may discover the nonarbitrary relationship between a number of objects and a number of sounds. Yet is such a return to constructivism plausible? Some objects generate more than one sound, others no sound at all. Environmental cues are therefore not devoid of ambiguity, and it is highly unclear that they would support any form of learning. I therefore suspect that the babies' preference for a correspondence between sounds and objects stems from an innate, abstract competence for numbers.

A similar uncertainty reigns over addition and subtraction abilities. Karen Wynn's 1+1 and 2–1 experiments have been performed only with babies who were four months and a half at youngest. This lapse of time may be sufficient for the baby to empirically discover that when one object and then a second disappear behind a screen, two objects will be found if one cares to look for them. In that case, Piaget would be partially right after all: Babies would have to extract the elementary rules of arithmetic from their environment—although they would do so at a much more precocious age than he imagined. Yet this knowledge may, rather, be inborn, built into the very architecture of the baby's brain, and become manifest as soon as the ability to memorize the presence of objects behind a screen emerges at around four months of age.

Whatever its origin, a rudimentary numerical accumulator clearly enables infants as early as six months of age to recognize small numbers of objects or sounds and to combine them in elementary additions and subtractions. Curiously, the one simple arithmetical notion that they may be lacking is the ordering of numbers. At what age do we know that 3 is larger than 2? Few experiments have studied this question in very young children, and none is really convincing. Yet their results suggest that no noticeable ordinal competence is found before the age of about fifteen months. At this age, children start to behave like the macaques Abel and Baker or the chimpanzee Sheba: They spontaneously select the larger of two sets of toys. Younger babies seem unaware of the natural ordering of numbers. It is as if their numerical detectors, programmed to respond to one, two, or three objects, entertained no particular relationship to one another. Perhaps we can liken the babies' representation of the numbers 1, 2, and 3 to our adult knowledge of the colors blue, yellow, and green. We can recognize these colors and we may even know how they combine ("blue plus yellow makes green"), yet we have absolutely no concept of an order in which to sort them. Likewise, babies can recognize one, two, or three objects and even know that 1 plus 1 makes 2, without necessarily realizing that 3 is larger than 2 or that 2 is larger than 1.

If these preliminary data can be trusted, then the concepts of "smaller" and "greater" are among the slowest to be put in place in the baby's mind. Where would they arise from? Probably from an observation of the properties of addition and subtraction. The "greater" number would be the number that you can reach by adding, and the "smaller" number the one that you can reach by subtracting. Babies would discover that the same relation "greater than" exists between 2 and 1 as between 3 and 2 because the same addition operation, "+1," enables one to move from 1 to 2 and from 2 to 3. By practicing successive additions, children would see the detectors for 1, 2, and 3 light up in a reproducible order in their mind and would thus learn about their position in the series of numbers.

But this is still a hypothetical scenario. A whole series of experiments would have to be performed before it could be confirmed or rejected. The one thing that we do know, at this stage, is that babies are much better mathematicians than we though only fifteen years ago. When they blow the first candle on their birthday cake, parents have every reason to be proud of them, for they have already acquired, whether by learning or by mere cerebral maturation, the rudiments of arithmetic and a surprisingly articulate "number sense."

Chapter 3

The Adult Number Line

I recommend you to question all your beliefs,
except that two and two make four.
Voltaire, *L'homme aux quarante écus*

I have long been intrigued by Roman numerals. There is something of a contradiction between the simplicity of the first numerals and the perplexing complexity of the others. The first three numerals, I, II, and III, follow a self-evident rule: They simply contain as many bars as there are units. Number IV, however, breaks the rule. It introduces a new sign, V, whose meaning is far from obvious, and a subtraction operation, 5–1, that seems arbitrary—why not 6–2, 7–3, or even 2×2?

Looking at the history of numerical notation, we find that the first three Roman numerals are like living fossils—they draw us back to a remote time when humans had not yet invented a way of writing down numbers and found it sufficient to keep track of numbers by engraving a stick with as many notches as the sheep or camels they owned. The series of notches preserved a durable record of a past accounting. This was indeed the very beginning of a symbolic notation, because the same row of five notches could symbolize any set of five objects.

This historical reminder, however, only thickens the mystery surrounding the fourth Roman numeral. Why did people abandon a notation that was so useful and simple? How did the arbitrariness of IV, which puts a burden on the attention and memory of the reader, come to replace the simplicity of IIII, which enabled the average shepherd to understand numbers? More to the point, if for one reason or another some revision of the number notation system was required, why did the first numerals I, II, and III escape it?

Is it just a historical accident? Some chance events must have presided over the fate of Roman number notation and its survival up to the present time. And yet the singularity of the Roman numerals I, II, and III has a universal character that transcends the mere history of Mediterranean countries. Georges Ifrah, in his comprehensive book on the history of numerical notations, shows that in *all* civilizations, the first three numbers were initially denoted by repeatedly writing down the symbol for "one" as many times as necessary, exactly as in Roman numerals. And most, if not all civilizations stopped using this system beyond the number 3 (see Figure 3.1). The Chinese, for instance, denote the numbers 1, 2, and 3 using one, two, and three horizontal bars—yet they employ a radically different symbol for number 4. Even our own Arabic digits, although they seem arbitrary, derive from the same principle. Our digit 1 is a single bar, and our digits 2 and 3 actually derive from two or three horizontal bars that became tied together when they were deformed by being handwritten. Only the Arabic digits 4 and beyond can thus be considered as genuinely arbitrary.

Dozens of human societies around the world have progressively converged on the same solution. Nearly all of them have agreed to denote the first three or four numbers by an identical number of marks, and the following numbers by essentially arbitrary symbols. Such a remarkable cross-cultural convergence calls for a general explanation. It seems clear enough that aligning nineteen marks to denote number 19 would impose an unbearable burden on number writing and

Cuneiform Notation	𒁹	𒈫	𒐈	𒐉	𒐊
Etruscan Notation	I	II	III	IIII	Λ
Roman Notation	I	II	III	IV	V
Mayan Notation	•	••	•••	••••	—
Chinese Notation	一	二	三	四	五
Ancient Indian Notation	—	=	≡	+	Y
Handwritten Arabic					
Modern "Arabic" Notation	1	2	3	4	5

Figure 3.1. Across the world, humans have always denoted the first three numbers by series of identical marks. Almost all civilizations abandon this analog notation beyond the numbers 3 or 4, which mark the limits of man's "immediate" apprehension of number. (Redrawn from Ifrah 1994.)

reading: Writing down nineteen strokes is a time-consuming and error-prone operation, and how could the reader possibly distinguish nineteen from eighteen or twenty? The emergence of number notations more compact than mere rows of bars therefore seemed inevitable. Yet this still does not explain why all nations have consistently elected to get rid of this system beyond the number 3 rather than, say, 5, 8, or 10.

At this point, it is tempting to draw a parallel with infants' number discrimination abilities. Human infants readily discriminate between one and two objects or between two and three objects, but their abilities do not extend much beyond this point. Obviously, infants do not contribute much to the evolution of number notations. Yet suppose that number discrimination abilities remained unchanged in human adults. This might provide the first elements of an explanation: Beyond number 3, the bar notation would no longer be legible because we would be unable to distinguish IIII from IIIII at a glance.

Roman numerals, then, lead us to examine to what extent the proto-numerical abilities found in animals and human babies extend to human adults. In this chapter, we hunt for living fossils and other cues such as Roman numerals which draw us back to the very foundations of human arithmetic. Indeed, we find many indications that the protonumerical representation of quantities still lives within us. Though mathematical language and culture have obviously enabled us to go way beyond the limits of the animal numerical representation, this primitive module still stands at the heart of our intuitions about numbers. It retains a considerable influence on our way of perceiving, conceiving, writing down, or speaking about numbers.

1, 2, 3, and Beyond

The fact that there is a strict limit on the number of objects that we are able to enumerate at once has been known to psychologists for at least a century. In 1886, James McKeen Cattel, in his laboratory at Leipzig, demonstrated that when subjects were briefly shown a card bearing several black dots, they could enumerate them with unfailing precision only if their number did not exceed three. Beyond this limit, errors accumulated. H. C. Warren, then at Princeton, and later Bertrand Bourdon, at the Sorbonne in Paris, each developed new methods of investigation to accurately measure the time required to quantify sets of objects. In 1908, Bourdon did not have any high-tech experimental equipment at his disposal. His experiments, most often performed on himself, involved the tinkering of specials tools. Let me quote from his original publication:

> The numbers, which were composed of horizontally aligned bright dots, were one meter away from my eyes. A sheet of copper with a rectangular opening, falling from a fixed height, let them be visible for a very short time.... To measure response times, I used a carefully adjusted Hipp chronoscope [an electro-mechanical chronometer accurate to within one thousandth of a second]. The electrical circuit through the chronoscope was closed when the dots started to become visible. Within this circuit was inserted a buccal switch which consisted for the most part in two separate copper leaves, one side of which was covered with fiber to insulate them from the mouth; I held these leaves between my teeth, clenching them so that the leaves would touch; then I would name the numbers as fast as possible as soon as I had recognized them, and for this purpose I had to unclench my teeth, which interrupted the circuit.

It was with this rudimentary apparatus that Bourdon discovered the fundamental law of visual quantification in humans. The time required to name a number of dots grows slowly from 1 to 3, and then suddenly increases sharply beyond this limit. At the very same point, the number of errors also jumps abruptly. This result, which has been replicated hundreds of times, remains valid to this day. It takes less than half a second to perceive the presence of one, two, or three objects. Beyond this limit, speed and accuracy fall dramatically (Figure 3.2).

A careful measurement of the response time curve reveals several important details. Between three and six dots, the increase in response time is *linear*, which means that it takes a fixed additional duration to enumerate each additional dot.

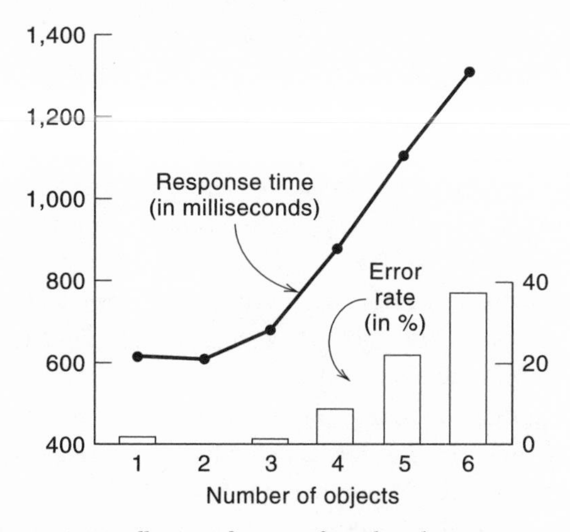

Figure 3.2. Enumerating a collection of items is fast when there are one, two or three items, but starts slowing down drastically beyond four. Errors begin to accumulate at the same point. (Redrawn from Mandler and Shebo 1982.)

It takes an adult about 200 or 300 milliseconds to identify each dot beyond three. This slope of 200 to 300 milliseconds corresponds roughly to the time it takes an adult to recite numbers when counting aloud as fast as possible. In children, the speed of reciting numbers drops to one number every one or two seconds—and the slope of the response time curve increases by the same amount. Thus to enumerate a set comprising more than three dots, adults and children alike have to count the dots at a relatively slow rate.

But then why is the enumeration of numbers 1, 2, and 3 so fast? The flattening of the response time curve within this region suggests that the first three dots do not have to be counted one by one. The numbers 1, 2, and 3 seem to be recognized without any appearance of counting.

While psychologists are still pondering over how such enumeration without counting might work, they have at least conceived of a name for it. It is called the "subitization" or "subitizing" ability, a name deriving from the Latin *subitus,* which means sudden. This is something of a misnomer since subitization, however fast, is anything but instantaneous. It takes about five- or six-tenths of a second to identify a set of three dots, or about the time it takes to read a word aloud or identify a familiar face. Neither is this duration constant: It slowly increases from 1 to 3. Hence, subitization probably requires a series of visual operations all the more complex the greater the number to be recognized.

What are these operations? A widely held theory supposes that we recognize small sets of one, two, or three objects rapidly because they form easily recognizable geometrical configurations: One object forms a dot, two make a line, and three, a triangle. This hypothesis, however, cannot explain the observation that we still subitize small sets whose objects are perfectly aligned, thus destroying all geometrical cues. Indeed, it is hard to see what geometrical parameters distinguish the Roman numerals II and III—yet we readily subitize them.

Psychologists Lana Trick and Zenon Pylyshyn, however, found a situation in which subitizing fails: when the objects are superimposed so that their locations are not readily perceptible. When viewing concentric circles, for instance, we have to count in order to determine whether there are two, three, or four of them. Thus, the subitizing procedure seems to require objects to occupy distinct locations—a cue that, as we saw earlier, is also exploited by babies to determine how many objects are present.

I therefore believe that subitizing in human adults, like numerosity discrimination in babies and animals, depends on circuits of our visual system that are dedicated to localizing and tracking objects in space. The occipito-parietal areas of the brain contain neuronal ensembles that rapidly extract, in parallel across the visual field, the locations of surrounding objects. Neurons in these areas seem to encode the location of objects regardless of their identity and even to maintain a representation of objects that have been hidden behind a screen. Hence, the information they extract is ideally abstract to feed an approximate accumulator. During subitizing, I believe that those areas quickly parse the visual scene into discrete objects. It is then easy enough to tally them up in order to obtain an estimate of their numerosity. The neural network simulation I developed with Jean-Pierre Changeux, which was described in Chapter 1, shows how this computation can be implemented by simple cerebral circuits.

Why would this mechanism introduce a discontinuity between 3 and 4? Remember that the accuracy of the accumulator decreases with numerosity. Hence it is increasingly difficult to distinguish a number n for its neighbors $n+1$ and $n-1$. Number 4 seems to be the first point where our accumulator starts to make a significant number of discrimination errors, confusing it with 3 or 5. This is why we have to count beyond the limit of 4—our accumulator still provides us with a numerosity estimate, but one that is no longer accurate enough to select a unique word for naming.

The theory of a "parallel accumulation of object locations" that I just sketched is not the only available theory of subitization, however. According to UCLA psychologists Randy Gallistel and Rochel Gelman, when we subitize, even if we are not aware of it, we always count the elements one by one—but very quickly. Subitizing would thus be a kind of fast serial counting without words. Although this seems counterintuitive, subitization would actually require the orienting of attention toward each object in turn and would therefore rely on a serial step-by-step algorithm. This is where the major testable difference with my hypothesis lies. My model suggests that, during subitizing, all the objects in the visual field are processed simultaneously and without requiring attention—what in cognitive psychologists' jargon is called "parallel preattentive processing". In my network simulation, number detectors start to respond at about the same time whether one, two, or three objects are present (although as the input numerosity gets larger they do take a slightly longer time to stabilize to the precise activation pattern that is needed for naming). Most importantly, in contrast to Gelman and Gallistel's fast counting hypothesis, my number detectors do not require each object to be singled out in turn by an mental "spotlight" or tagging process—all are taken in at once and in parallel.

Although the jury is still out on this issue, perhaps the best evidence that subitizing does not require serial orienting of attention comes from human patients who, following a cerebral lesion, are unable to attentively explore their visual environment and therefore to count. Mrs. I, whom I have examined together with Dr. Laurent Cohen at the hôpital de la Salpêtrière in Paris, suffered from a posterior cerebral infarct due to high blood pressure during her pregnancy. One year later, the after-effects of this lesion on her visual perception abilities were still present. Mrs. I had become unable to recognize certain visual shapes, including faces, and she also complained of curious distortions of her vision. When we asked her to describe a complex image, she often omitted important details and did not perceive the meaning of the whole. Neurologists call this deficit "simultanagnosia." It made counting impossible for her. When four, five, or six dots were briefly flashed on a computer screen, she almost always forgot to count some of them. She attempted to count but failed to orient toward each object in turn. Once she had counted about half of the items, she stopped because she thought that she had counted them all. Another patient with a similar deficit sank into an opposite pattern of error: She failed to take good note of the items that she had already counted, and she kept on counting the same items over and over again. She would tell us, without batting an eyelid, that there were twelve points when in fact there were only four!

Despite their terrible counting handicap, however, these two patients experienced astonishingly little difficulty in enumerating sets of one, two, or even three dots. With small numbers, they responded quite rapidly, confidently, and almost always flawlessly. Mrs. I, for instance, made errors only 8% of the time when enumerating three items, but she erred 75% of the time when enumerating four items. This dissociation is one we have often observed: The perception of small numerosities can remain intact even though a cerebral lesion makes it totally impossible for the patient to sequentially orient attention toward each object in turn. This strongly suggests that subitizing does not involve sequential counting, but merely a parallel and preattentive extraction of objects in the image.

Approximating Large Numbers

In the motion picture *Rain Man*, in which Dustin Hoffman plays Raymond, an autistic man with prodigious abilities, a peculiar event occurs. A waitress drops a box of toothpicks on the floor, and Raymond immediately utters "82 . . . 82 . . . 82 . . . that's 246!" as if he had counted the toothpicks by groups of 82 in less time than it would take us to say "2 and 2, 4." In Chapter 6, we will analyze in detail the feats that have been attributed to calculating prodigies such as Raymond. Let me say right now, however, that in this particular case, I do not believe that Dustin Hoffman's performance should be taken at face value. A few anecdotal reports have been made of fast enumeration in some autistic patients; but there have been no response time measurements that I know of that might help determine whether these people do indeed count. My own experience is that simulating Rain Man's performance is relatively easy by starting to count in advance, by mentally adding groups of dots, and by bluffing a bit. (Just one success at guessing the exact number of people in a room is often sufficient to turn you into a legend!) The most likely possibility, then, is that the subitizing limit of three or four items applies equally to all humans.

But what is the nature of this limit? Are our parallel enumeration abilities really paralyzed when a set comprises more than three items? Do we necessarily have to count when this limit is reached? In fact, any adult can estimate, within a reasonable margin of uncertainty, numbers way beyond 3 or 4. The subitizing limit is therefore not an insurmountable barrier, but a mere borderline beyond which there is a universe of approximation. When confronted with a crowd, we may not know whether there are eighty-one, eighty-two, or eighty-three people, but we can estimate eighty or one hundred without counting.

Such approximations are generally valid. Psychologists do know of situations in which human estimations systematically deviate from the real value (Figure 3.3). For instance, we all tend to overestimate numerosity when the objects are regularly spread out on a page, and conversely we tend to underestimate sets of irregularly distributed objects, perhaps because our visual system parses them into small groups. Our estimations are also sensitive to context, leading us to underestimate or overestimate the very same set of thirty dots depending on whether it is surrounded by sets of ten or of one hundred dots. As a rule, however, our approximations are remarkably accurate, especially considering the rarity of occasions in which we can verify their correctness in everyday life. How often indeed do we get exact feedback as to whether a crowd is made up of one hundred, two hundred, or five hundred people? Yet in a laboratory experiment, it has been shown that one single exposure to veridical numerical information, such as a set of two hundred dots dutifully labeled as such, suffices to improve our estimations of sets of between ten and four hundred dots. To calibrate our number estimation system, only a handful of precise measurements are required.

Far from being exceptional, our perception of large numbers follows laws that are strictly identical to those that govern animal numerical behavior. We are subject to a distance effect: We more easily distinguish two distant numerosities, such as 80 and 100, than two closer numbers such as 81 and 82. Our perception of numerosity also exhibits a magnitude effect: For an equal distance, we have a harder time discriminating two large numerosities, such as 90 and 100, than two small ones, such as 10 and 20.

These laws are remarkable for their unfailing mathematical regularity, an unusual finding in psychology. Suppose that a given person can discriminate, with an accuracy of 90%, a set of thirteen dots from another reference set of ten

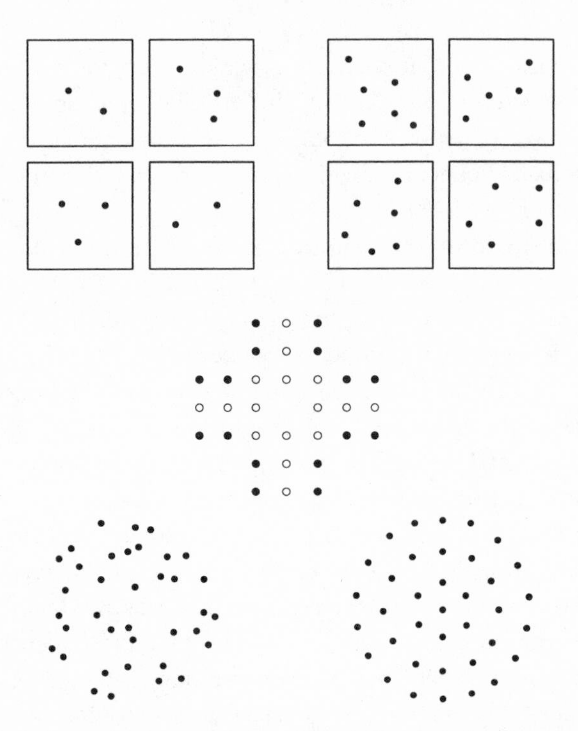

Figure 3.3. The difference between 2 and 3 items (top left) is immediately perceptible to us, but we cannot distinguish 5 from 6 (top right) without counting. Our perception of large numbers relies on the density of items, the area they occupy, and the regularity of their distribution in space. In the middle "solitaire illusion," first described by Uta and Christopher Frith in 1972, our perceptual apparatus incorrectly convinces us that there are more white dots than black dots, probably because the white dots are more tightly grouped. Bottom, randomly distributed dots seem less numerous than regularly spaced ones; each disk actually has 37 dots.

dots (hence a numerical distance of 3). Let us now double the size of the reference set to twenty dots. How far from this numerosity do we have to move to again reach 90% correct discrimination? The answer is quite simple: one merely has to double the numerical distance to 6, and hence to present a set of twenty-six dots. When the reference number doubles, so does the numerical distance that humans can discriminate within a fixed level of performance. This multiplication principle is also known as the "scalar law" or "Weber's law," after the German psychologist who discovered it. Its remarkable similarity to the laws that govern animal behavior proves that, inasmuch as the approximate perception of numerosity is concerned, humans are no different from rats or pigeons. All our mathematical talent is useless when it comes to perceiving and estimating a large number.

The Quantity Behind the Symbols

That our apprehension of numerosity does not differ much from that of other animals may seem unremarkable. After all, mammals share a fundamentally similar visual and auditory perception apparatus. In some domains, such as olfaction, human perceptual abilities even turn out to be quite inferior to those of other species. But when it comes to language, one might think that our performance should set us apart from the rest of the animal kingdom. Obviously, what distinguishes us from other animals is our ability to use arbitrary symbols for numbers, such as words or Arabic digits. These symbols consist of discrete elements that can be manipulated in a purely formal way, without any fuzziness. Introspection suggests that we can mentally represent the meaning of numbers 1 through 9 with equal acuity. Indeed, these symbols seem equivalent to us. They all seem equally easy to work with, and we feel that we can add or compare any two digits in a small and fixed amount of time, like a computer. In summary, the invention of numerical symbols should have freed us from the fuzziness of the quantitative representation of numbers.

How misleading these intuitions can be! Although numerical symbols have provided us with a unique door to the otherwise inaccessible realms of rigorous arithmetic, they have not severed our roots with the approximate animal representation of quantities. Quite to the contrary, each time we are confronted with an Arabic numeral, our brain cannot but treat it as a analogical quantity and represent it mentally with decreasing precision, pretty much as a rat or chimpanzee would do. This translation from symbols to quantities imposes an important and measurable cost to the speed of our mental operations.

The first demonstration of this phenomenon dates back to 1967. At that time, it was judged so revolutionary as to deserve the honors of a publication in the journal *Nature*. Robert Moyer and Thomas Landauer had measured the precise time an adult took to decide which of two Arabic digits was the largest. Their experiment consisted in flashing pairs of digits such as 9 and 7 and asking the subject to report where the larger digit was located by pressing one of two response keys.

This elementary comparison task was not as easy as it appeared. The adults often took more than half a second to complete it, and the results were not error-free. Even more surprising, performance varied systematically with the numbers chosen for the pair. When the two digits stood for very different quantities such as 2 and 9, subjects responded quickly and accurately. But their response time slowed by more than 100 milliseconds when the two digits were numerically closer, such as 5 and 6, and subjects then erred as often as once in every ten trials. Moreover, for equal distance, responses also slowed down as the numbers became increasingly larger. It was easy to select the larger of the two digits 1 and 2, a little harder to compare digits 2 and 3, and far harder to respond to the pair 8 and 9.

Let there be no misunderstanding: The people that Moyer and Landauer tested were not abnormal, but individuals like you and me. After experimenting on number comparison for more than ten years, I still have yet to find a single subject who compares 5 and 6 as quickly as he or she compares 2 and 9, without showing a distance effect. I once tested a group of brilliant young scientists, including students from the top two mathematical colleges in France, the Ecole Normale Supérieure and the Ecole Polytechnique. All were fascinated to discover that they slowed down and made errors when attempting to decide whether 8 or 9 was the larger.

Nor does systematic training help. In a recent experiment, I attempted to train some University of Oregon students to escape the distance effect. I simplified the task as much as possible by presenting only the digits 1, 4, 6, and 9 on a computer screen. The students had to press a right-hand key if the digit they saw was larger than 5, and a left-hand key if it was smaller than 5. One can hardly think of a simpler situation: If you see a 1 or a 4, press left, and if you see a 6 or 9, press right. Yet even after several days and 1,600 training trials, the subjects were still slower and less accurate with digits 4 and 6, which are close to 5, than with digits 1 and 9, which are further away from 5. In fact, although the responses became globally faster in the course of training, the distance effect itself—the difference between digits close to 5 and far from 5—was left totally unaffected by training.

How are we to interpret these number comparison results? Clearly, our memory does not preserve a stored list of responses for all possible digit comparisons. Were we to learn all possible combinations of digits by rote—for instance, that 1 is smaller than 2, 7 larger than 5, and so on—comparison times should not vary with number distance. Where, then, does the distance effect come from? As far as physical appearance is concerned, digits 4 and 5 are no more similar than digits 1 and 5. Hence the difficulty in deciding whether 4 is smaller or larger than 5 has nothing to do with a putative difficulty in recognizing the shapes of digits. Obviously, the brain does not stop at recognizing digit shapes. It rapidly recognizes that at the level of their *quantitative meaning*, digit 4 is indeed closer to 5 than 1 is. An analogical representation of the quantitative properties of Arabic numerals, which preserves the proximity relations between them, is hidden somewhere in our cerebral sulci and gyri. Whenever we see a digit, its quantitative representation is immediately retrieved and leads to greater confusion over nearby numbers.

One more striking demonstration of this fact is what occurs when we compare two-digit numerals. Suppose you had to decide whether 71 was smaller or larger than 65. One rational approach is to initially examine only their leftmost digits, 7 and 6, to note that 7 is larger than 6, and to conclude that 71 is larger than 65 without even considering the identity of the rightmost digits. Indeed, this sort of algorithm is used by computers to compare numbers. But this is not how the human brain does it. When one measures the time it takes to compare several two-digit numbers with 65, a smooth continuous curve is found (Figure 3.4). Comparison time increases continuously as the numbers to be compared become increasingly close to the reference number 65. Both the left and the right digits contribute to this progressive increase. Thus, it takes more time to figure out that 71 is larger than 65 than to reach the same decision for 79 and 65, although the leftmost digit 7 is the same in both cases. Furthermore, responses are not disproportionately slowed when the decades change: Comparing 69 with 65 is just a bit slower than comparing 71 with 65, whereas it should be much more difficult if we were indeed selectively attending initially to the leftmost digit only.

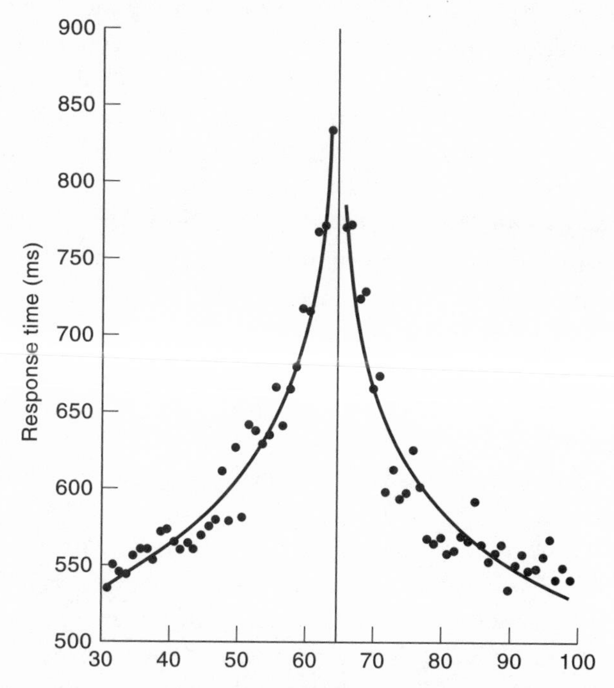

Figure 3.4. How long does it take to compare two numbers? Thirty-five adult volunteers classified all two-digit Arabic numerals between 31 and 99 as being smaller or larger than 65, while their responses were timed to the nearest millisecond. Each black dot shows the average response time to a given number. Responses become increasingly slow as the target numeral gets closer to 65: the distance effect. *(Data from Dehaene et al. 1990.)*

The only explanation I can come up with is that our brain apprehends a two-digit numeral as a whole and transforms it mentally into an internal quantity or magnitude. At this stage, it forgets about the precise digits that led to this quantity. The comparison operation is concerned only with numerical quantities, not the symbols that convey them.

The Mental Compression of Large Numbers

The speed with which we compare two Arabic numerals does not depend solely on the distance between them, but also on their size. It takes much more time to decide that 9 is larger than 8 than to decide that 2 is larger than 1. For equal distance, larger numbers are more difficult to compare than smaller ones. This slowing down for large numbers is again reminiscent of the perceptual abilities of babies and animals, which are similarly affected by numerical distance and size effects. Such an astonishing parallel confirms that, starting with a symbol such as an Arabic numeral, our brain retrieves an internal representation of quantities remarkably similar to the one present in animals and young children.

In fact, just as in animals, the parameter that governs the ease with which we distinguish two numbers is not so much their absolute numerical distance but their distance relative to their size. Subjectively speaking, the distance between 8 and 9 is not identical to that between 1 and 2. The "mental ruler" with which we measure numbers is not graduated with regularly spaced marks. It tends to compress larger numbers into a smaller space. Our brain represents quantities in a fashion not unlike the logarithmic scale on a slide rule, where equal space is allocated to the interval between 1 and 2, between 2 and 4, or between 4 and 8. As a result, the accuracy and speed with which calculations can be performed necessarily decreases as the numbers get larger.

Many an empirical result may be summoned to support the hypothesis of the mental compression of large numbers. Some experiments are based solely on introspection. What number subjectively rates as being closer to 5: 4 or 6? Although the question seems farfetched, most people respond that for equal distance, the larger number 6 seems to differ less. Other experiments have used more subtle and indirect methods. For instance, let us pretend that you are a random number generator and that you have to select numbers at random between 1 and 50. Once this experiment is performed on a large number of subjects, a systematic bias emerges: Instead of responding randomly, we tend to produce smaller numbers more frequently than larger ones—as if smaller numbers were overrepresented in the "mental urn" from which we were drawing. This should persuade us to never draw anything at random without relying on an "objective" source of randomness, such as dice or a real random number generator!

I suspect that this bias for small numbers has far-reaching and sometimes pernicious consequences for the way we use our intuition to conduct and interpret statistical analyses. Consider the following problem. Two series of numbers have been generated at random by a computer. Without making any calculations, your task is to rate how randomly and evenly each series seems to sample the interval of numbers between 1 and 2,000:

Series A:	879	5	1,322	1,987	212	1,776	1,561	437	1,098	663
Series B:	238	5	689	1,987	16	1,446	1,018	58	421	117

Most people respond that the numbers in series B are more evenly spread out and therefore "more random" than those in series A. In series A, large numbers seem to appear much too often. And yet from a mathematical point of view, it is A, and not B, that samples the continuum of numbers between 1 and 2,000 best. The numbers in series A are regularly spaced by just over 200 units, whereas those in series B are exponentially distributed. The reason why we prefer series B is that it fits best with our mental idea of the number line, which is pictured as a compressed series in which larger numbers are less conspicuous than smaller ones.

A compression effect is also perceptible in the way we select units of measurement. On April 17, 1795, of the French republic—Germinal 18th, year III, of the "revolutionary calendar"—the metric system was instituted in Paris. Aiming at universality, its units covered a whole range of powers of ten, from nanometer to kilometer. Even though each power of ten received a specific name—millimeter, centimeter, decimeter, meter, and so on—these units were still spaced too far apart to be practical for everyday use. So the French lawmakers stipulated that "each decimal unit shall have its double and its half". From this stipulation derived the regular series 1, 2, 5, 10, 20, 50, 100 . . . , still in use today for coins and banknotes. It fits our number sense because it approaches an exponential series while comprising only small round numbers. In 1877, similar constraints led Colonel Charles Renard to adopt a method for the normalizing of industrial products such as bolt diameters or wheel sizes that was based on another quasi-logarithmic series (100, 125, 160, 200, 250, 315, 400, 500, 630, 800, 1,000). As soon as a continuum needs to be divided into discrete categories, intuition dictates the selection of a compressed scale, most often logarithmic, which tightly matches our internal representation of numbers.

Reflexive Access to Number Meaning

An Arabic numeral first appears to us as a distribution of photons on the retina, a pattern identified by visual areas of the brain as being the shape of a familiar digit. Yet the many examples that we have just described show that the brain hardly pauses at recognizing digit shapes. It rapidly reconstructs a continuous and compressed representation of the associated quantity. This conversion into a quantity occurs unconsciously, automatically, and at great speed. It is virtually impossible to see the shape of digit 5 without immediately translating it into quantity five—even when this translation is of no use at all in the current context. Understanding numbers, then, occurs as a reflex.

Suppose you were shown two digits side by side and were asked to tell, as fast as you can, whether they were the same or different. Surely you'd think that you might base your decision exclusively on the visual appearance of the digits—whether or not they share the same shape. But measurement of response times shows that this supposition is wrong. Deciding that 8 and 9 are different digits takes systematically longer than reaching the same decision for digits 2 and 9. Once again, numerical distance governs our speed of responding. Quite unconsciously, we are reluctant to respond that 8 and 9 are different digits because the quantities that they represent are so similar.

A similar "comprehension reflex" also affects our memory for digits. Memorize the following list of digits: 6, 9, 7, 8. Done? Now tell me whether digit 5 figured in the list. And what about digit 1? Does the first question seem harder than the second? Although the correct response is "no" in both cases, formal experiments show that the more distant the probe digit is from the memorized list, the shorter the response time. The list is obviously not memorized only as a series of arbitrary symbols, but also as a swarm of quantities close to 7 or 8—which is why we can immediately tell that 1 is not in the set.

Is it ever possible to inhibit the comprehension reflex? To answer this question, subjects can be placed in a situation where it is really advantageous for them *not to know* the meaning of digits. Two Israeli researchers, Avishai Henik and Joseph Tzelgov, presented pairs of digits of different sizes such as 1 and 9 on a computer screen. They measured how much time subjects required to indicate the symbol that was printed in larger font. This task requires subjects to focus their attention on physical size and to neglect, as much as possible, the numerical size of the digits. Once again, however, an analysis of response times shows how automatic and irrepressible the comprehension of numerals is. It is much easier for subjects to respond when the physical and numerical dimensions of the stimuli are congruent, as in the pair 1 9, than when they are conflicting, as in the pair

9 1. We apparently cannot forget that the symbol "1" means a quantity smaller than nine.

Even more surprisingly, access to numerical quantity can occur in our brain under conditions in which we are not even aware of having seen a digit. By presenting a symbol on a computer for a very short period of time, it can be made to appear invisible. One technique that psychologists call "sandwich priming" consists in preceding and following the word or digit one desires to hide by a meaningless character string. One may, for instance, show "#######," then the word "five," then "#######," and finally the word "SIX". If the first three strings are each presented only for one twentieth of a second, the prime word "five," sandwiched between the other strings, becomes invisible—not just difficult to read, but vanished from the stream of consciousness. Under the right conditions, even the programmer of the experiment cannot tell whether the hidden word is present or not! Only the first string "#######" and the word "SIX" remain consciously visible. Yet for 50 milliseconds, a perfectly normal visual stimulus "five" was present on the retina. In fact unbeknownst to the subject, it even contacted a whole series of mental representations in his or her brain. This can be proved by measuring the time taken to name the target word "SIX": It varies systematically with the numerical distance between the prime word and the target word. Naming the word "SIX" is faster when it is preceded by a close prime such as "five" than when it is preceded by a more distant prime such as "two." Hence, the comprehension reflex unfolds in this situation too: Although the word "five" has not been consciously seen, it is still interpreted by the brain as "a quantity close to six."

Although we are not aware of all the automatic numerical computations that are continuously being handled in our brain circuits, their impact in our daily lives is certain and can be illustrated in numerous ways. In a major train station in Paris, the platforms are numbered, but the design of the station, which is divided into several distinct zones, imposes a disruption in the number sequence: Platform 11 is next to platform 12, but platform 13 is far away. So deeply is the continuity of numerical quantities engraved in our minds that this design throws many travelers into disarray. Our intuition imposes that platform 13 be next to platform 12.

Along the same lines, here is a factoid guaranteed to catch your attention:

> "St. Theresa of Àvila died during the night between the 4th and 15th of October 1582."

No, this is not a typo! As luck would have it, the saint died on the very night on which pope Gregory XIII abrogated the ancient Julian calendar, instituted by

Julius Caesar, and replaced it with the Gregorian calendar still in use. The adjustment, which was made necessary by the progressive shift of calendar dates from astronomical events such as solstices over the course of centuries, deemed that the day after October 4 became October 15—a punctual decision, but one that profoundly upsets our sense of the continuity of numbers.

The automatic interpretation of numbers is also exploited in the field of advertising. If so many retailers take the trouble to mark price tags at $399 instead of $400, it is because they know that their clients will automatically think of this price as being "about 300 dollars"; only on reflection will they realize that the actual sum is very close to 400 dollars.

As a final example, let me report on my own experience of having to adapt to the Fahrenheit temperature scale. In France, where I was born and raised, we use only the centigrade scale, in which water freezes at 0° and boils at 100°. Even after living in the United States for two years, I still found it difficult to think of 32°F as cold, because for me 32° automatically evoked the normal temperature on a very warm sunny day! Conversely, I suppose that most Americans traveling in Europe are shocked by the idea that anything as small as 37° can represent the temperature of the human body. The automatic attribution of meaning to numerical quantities is deeply embedded in our brains, and an adult can revise it only with great difficulty.

A Sense of Space

Numbers do not just evoke a sense of quantity; they also elicit an irrepressible feeling of extension in space. This intimate link between numbers and space was apparent in my number comparison experiments. As you may remember, subjects had to classify numbers as smaller or larger than 65. To this end they held two response keys, one in the left hand and the other in the right hand. Being a rather obsessive experimenter, I systematically varied the side of response: Half of the subjects responded "larger" with their right hand and "smaller" with their left hand, while the other group of subjects followed the opposite instructions. Surprisingly, this seemingly innocuous variable had an important effect: Subjects in the "larger-right" group responded faster and made fewer errors than those in the "larger-left" group. When the target number was larger than 65, subjects pressed the right-hand key faster than the left-hand key; the opposite was true for numbers smaller than 65. It was as if, in the subject's mind, large numbers were spontaneously associated with the right-hand side of space and small numbers with the left-hand side.

To what extent this association was automatic, it remained to be seen. To figure this out, I used a task that had little to do either with space or with quantity: Subjects now determined whether a digit was odd or even. Subsequently, other researchers have used even more arbitrary instructions such as discriminating whether a digit's name starts with a consonant or a vowel, or whether it has a symmetrical visual shape. Regardless of instructions, the same effect occurs: The larger the number, the faster right-hand responses are, compared with left-hand responses. And conversely, the smaller the number, the greater the bias toward responding faster on the left. As a tribute to Lewis Carroll, I called this finding the SNARC effect—an acronym for "Spatial-Numerical Association of Response Codes." (Carroll's wonderfully non-sensical poem, "The Hunting of the Snark," tells of the relentless quest for a mythical creature, the Snark, that no one has ever seen but whose behavior is known in exquisite detail, including its habit of getting up late and its fondness for bathing-machines—a very appropriate metaphor for scientists' obstinate pursuit of ever more accurate descriptions of nature, be they termed quarks, black holes, or universal grammars. Unfortunately, I could not think of a meaningful acronym for Carroll's original spelling of Snark!) The fact that the SNARC effect occurs whenever a digit is seen, even when the task itself is nonnumerical, confirms that it reflects the automatic activation of quantity information in the subject's brain.

Across the many experiments in which my colleagues and I "hunted the SNARC," we made a number of interesting discoveries. First, the absolute size of the numbers does not matter. What counts is their size relative to the interval of numbers used in the experiment. Numbers 4 and 5, for instance, are preferentially associated with the right if the experiment comprises only numbers from 0 to 5, and with the left if only numbers from 4 to 9 are used. Second, the hand used for responding is also irrelevant: When subjects respond while crossing the hands, it is still the right-hand side of *space* that is associated with larger numbers, even though right-sided responses are made using the left hand. And of course subjects are completely unaware of responding faster on one side than on the other.

The finding of an automatic association between numbers and space leads to a simple yet remarkably powerful metaphor for the mental representation of numerical quantities: that of a number line. It is as if numbers were mentally aligned on a segment, with each location corresponding to a certain quantity. Close numbers are represented at adjoining locations. No wonder, then, that we tend to confound them, as reflected by the numerical distance effect. Furthermore, the line can be metaphorically thought of as being oriented in space: Zero is at the extreme left, with larger numbers extending toward the right. This is why the reflex encoding of Arabic numerals as quantities is also accompanied by an automatic orientation of numbers in space, small ones to the left and large ones to the right.

What is the origin of this privileged axis oriented from left to right? Is it linked to a biological parameter such as handedness or hemispheric specialization, or does it depend only on cultural variables? Exploring the first hypothesis, I tested a group of left-handers—but they did not differ from right-handers and still associated large numbers with the right. Turning then to the second hypothesis, my colleagues and I recruited a group of twenty Iranian students who had initially learned to read from right to left, contrary to our Occidental tradition. This time, the results were more conclusive. As a group, Iranians did not show any preferential association between numbers and space. In each individual, however, the direction of the association varied as a function of exposure to Western culture. Iranian students who had lived in France for long showed a SNARC effect just like that of native French students, while those who had emigrated from Iran only a few years before tended to associate large numbers with the *left-hand* side of space rather than the right-hand side. Thus, it seems that cultural immersion is a major factor. The direction of the association between numbers and space seems to be related to the direction of writing.

A minute of reflection shows that indeed, the organization of our Western writing system has pervasive consequences on our everyday use of numbers. Whenever we write down a series of numbers, small numbers appear first in the sequence and hence to the left. In this way, left-to-right organization is imposed on rulers, calendars, mathematical diagrams, library bookshelves, floor signals above elevator doors, computer keyboards, and so on The internalization of this convention starts in childhood: Young American children already explore sets of objects from left to right, while Israeli children do the opposite. When counting, Occidental children almost always start on the left. The regular association of the beginning and ending points of counting with different directions of space then becomes internalized as an integral characteristic of the mental representation of number.

When this implicit convention is violated, we suddenly become painfully aware of its importance. Travelers entering Terminal 2 of the Charles de Gaulle airport in Paris experience a confusing situation: The gates bearing small numbers extend to the right, while those bearing large numbers extend to the left. I have observed many travelers, including myself, heading in the wrong direction after being assigned a gate number—a spatial disorientation that even repeated visits does not fully dissipate.

Although this had not yet been studied empirically, numbers are also probably associated with the vertical axis. I once stayed with colleagues at a hotel hanging from a cliff above the Adriatic Sea near Trieste, in Italy. The entrance was on the top floor, and perhaps for this reason successive floors were numbered from top to bottom. Confusion was always great when we took the elevator. Going up, we unconsciously expected the lighted floor numbers to increase, but the opposite occurred, perplexing us for a few seconds. We even had trouble figuring out which button to press to go one floor up! My hope is that architects and ergonomists, if they ever read this book, will adopt in the future a systematic rule of numbering from left to right and from bottom to top, for this is indeed a convention that our brains have come to expect, at least in our Western culture.

Do Numbers Have Colors?

Though a majority of people have an unconscious mental number line oriented from left to right, some have a much more vivid image of numbers. Between 5% and 10% of humanity is thoroughly convinced that numbers have colors and occupy very precise locations in space. In the 1880s already, Sir John Galton remarked that several acquaintances, most of them women, gave numbers extraordinarily precise and vivid qualities that were incomprehensible to anybody else. One of them described numbers as a ribbon undulating rightward, richly colored in shades of blue, yellow, and red (Figure 3.5). Another claimed that numbers from 1 to 12 coiled in a vaguely circular curve, with a slight break between 10 and 11. Beyond 12, the curve took off toward the left with distinct curls for each decade. A third person maintained that numerals from 1 to 30 appeared written in a vertical column in his mind's eye, and that the following decades progressively shifted to the right. According to him the numerals were "about half an inch long, of a light grey colour on a darker brownish grey colour."

Such "number forms," however outlandish, were not just inventions springing from the fertile minds of Victorians eager to please Galton's passion for numbers. A recent survey, conducted a century after Galton's, found similar images of numbers in modern university students—the same curves in some, the same straight lines in others, the same abrupt changes around decade boundaries, and so on. Furthermore, associations between numbers and colors are systematic: Most people associate black and white with either 0 and 1 or 8 and 9; yellow, red, and blue with small numbers such as 2, 3, and 4; and brown, purple, and gray with larger numbers such as 6, 7, and 8.

These statistical regularities suggest that most people who claim to experience number forms are sincere. They seem to faithfully describe a genuine percept, which can be extremely precise. One such person was given fifty colored pencils in order to couch her images of numbers on paper. On two different occasions

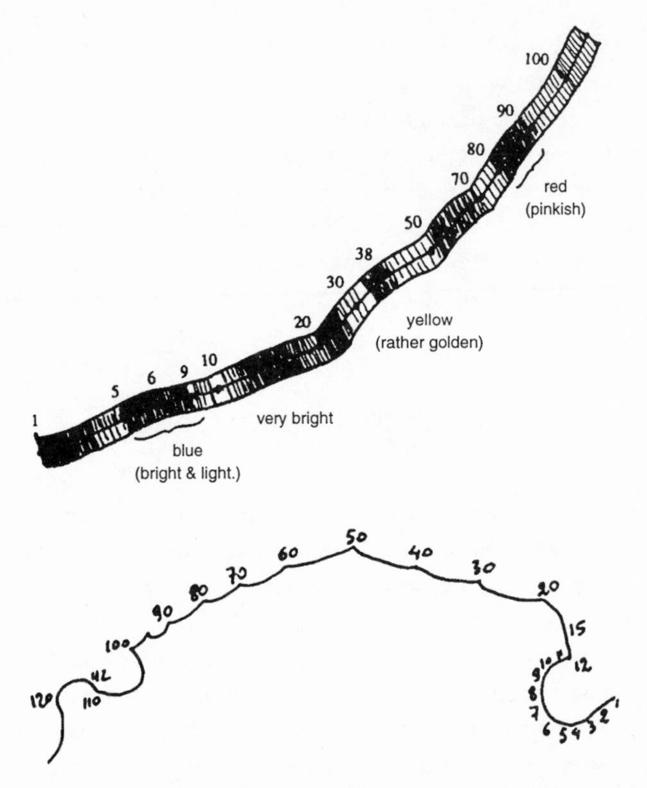

Figure 3.5. *These drawings describe the "number forms" experienced by two of Galton's subjects. One of them sees a colorful ribbon extending rightward. The second places numbers on a twisting curve whose initial section resembles the face of a clock. (Reprinted from Galton 1880 by permission of the publisher; copyright © 1880 by Macmillan Magazines Ltd.)*

separated by one week, she selected almost exactly the same shades of color. For some numbers, she even felt the need to mix the hues of several pencils to better depict her exact mental image.

Despite their rarity and strangeness, number forms share many properties with the "normal" representation of numerical quantities. The series of integers is almost always represented by a continuous curve, 1 falling next to 2, 2 next to 3, and so on. Only occasionally does one find abrupt changes in direction or small discontinuities at decade boundaries—for instance, between 29 and 30. Not a single person has yet claimed to see a jumbled image of numbers in which, say, primes or squares are grouped together on the same curve. The continuity of numerical quantities is the major parameter along which number forms are organized.

Relations between numbers and space are also respected. In most number forms, increasing numbers extend toward the top right. Finally, most people claim that their number form becomes increasingly fuzzy for larger numbers. This is reminiscent of the magnitude or compression effect that characterizes animal and human numerical behavior and limits the accuracy with which we can mentally represent large numbers.

In essence, then, number forms can be likened to a conscious and enriched version of the mental number line that we all share. While most people's mental number line is apparent only in subtle reaction time experiments, number forms are readily available to awareness and are also richer in visual details such as color or a precise orientation in space. Where do these illuminations come from? When questioned, the bearers of number forms either claim that they emerged spontaneously before the age of eight, or that they have had them for as long as they can remember. Sometimes several members of a family share the same type of number form. Yet this does not necessarily mean that a common genetic component is involved: The familial environment could also be a determinant.

My own speculation leads me to suppose that number forms may have something to do with how cortical maps of space and number are formed during development. As we have seen in chapter 2, babies may already possess a "mental map" of numerosity. Between the ages of three and eight, with schooling, the initial number line must be considerably enriched in order to accommodate the child's increasing knowledge of large numbers and of numeration in base 10. One might speculate that the acquisition of arithmetic is accompanied by a progressive expansion of the amount of cortex dedicated to the "number map" (such increases have indeed been observed in sensorimotor brain areas when an animal learns a fine manual task). As we shall see in Chapters 7 and 8, the inferior parietal cortex, a lateral and posterior brain region nearing the junction between the parietal, occipital, and temporal lobes, is a plausible candidate for where in the brain this expansion of the neural networks for arithmetical knowledge might occur. Because the total number of neurons remains constant, the growth of the numerical network must occur at the expense of the surrounding cortical maps, including those coding for color, form, and location. In some children, perhaps the shrinkage of nonnumerical areas may not reach its fullest term. In this case, some overlap between the cortical areas coding for numbers, space, and color may remain. Subjectively, this might translate into an irrepressible sensation of "seeing" the color and location of numbers. A similar account might explain the related phenomenon of synesthesia—the impression, familiar to poets or musicians, that sounds have shapes and that tastes evoke colors.

Speculative as it may be, this theory of how the cortex gets colonized by an increasingly refined map of numbers has some evidence in its favor. The neuropsychologists J. Spalding and Oliver Zangwill have described a twenty-four-year-old patient whose visual image of numbers disappeared suddenly when he experienced a lesion in the left parieto-occipital area, a region that has long been suspected to play a central role in mental arithmetic. Indeed, the patient suffered from severe difficulties both in calculating and in orienting in space (this neurological syndrome is discussed in more detail in Chapter 7). Hence, this case confirms that the subjective feeling of "seeing numbers" rests on the simultaneous coding of numerical and spatial information side by side in the same cerebral region.

Further, the idea that cortical maps may overlap and engender strange subjective sensations has been validated in studies of amputees. Following amputation of one arm, the region of the somatosensory cortex that represented this arm becomes vacant and is colonized by surrounding representations such as the head. In rare cases, it is then possible, by stimulating certain points of the face, to create sensations that feel as if they are coming from the missing arm, thus giving patients an irresistible impression of possessing a "phantom limb." A drop of water dripping on the face, for instance, feels as if the nonexistent arm were immersed in a bucket! I believe that the phenomenon of number forms, in which numbers evoke phantom colors and shapes, has a similar origin in overlapping cortical maps.

Intuitions of Number

It is now time to recapitulate the essential message of this chapter. These observations on Roman numerals, on the time it takes to compare Arabic numerals, and on some people's bizarre numerical hallucinations shed light on the fascinating peculiarities of our mental representation of numbers. An organ specialized in the perception and representation of numerical quantities lies anchored in our brains. Its characteristics unequivocally connect it to the protonumerical abilities found in animals and in infants. It can accurately code only sets whose numerosity does not exceed 3, and it tends to confuse numbers as they get larger and closer. It also tends to associate the range of numerical quantities with a spatial map, thus legitimizing the metaphor of a mental number line oriented in space.

Obviously, compared to babies and animals, human adults have the advantage of being able to convey numbers using words and digits. We will see in the next chapters how language eases the computation and communication of precise numerical quantities. However, the availability of precise number notations does not obliterate the continuous and approximate representation of quantities with which we are endowed. Much to the contrary, experiments show that the adult human brain, whenever it is presented with a numeral, rushes to convert it into an internal analogical magnitude that preserves the proximity relations between quantities. This conversion is automatic and unconscious. It allows us to retrieve immediately the meaning of a symbol such as 8—a quantity between 7 and 9, closer to 10 than to 2, and so on.

A quantitative representation, inherited from our evolutionary past, underlies our intuitive understanding of numbers. If we did not already possess some internal non-verbal representation of the quantity "eight," we would probably be unable to attribute a meaning to the digit 8. We would then be reduced to purely formal manipulations of digital symbols, in exactly the same way that a computer follows an algorithm without ever understanding its meaning.

The number line that we use to represent quantities clearly supports a limited form of intuition about numbers. It encodes only positive integers and their proximity relations. Perhaps this is the reason not only for our intuitive grasp of the meaning of integers, but also for our lack of intuition concerning other types of numbers. What modern mathematicians call numbers includes zero, negative integers, fractions, irrational numbers such as π, and complex numbers such as $i = \sqrt{-1}$. Yet all of these entities, except perhaps the simplest fractions such as ½ or ¼, posed extraordinary conceptual difficulties to mathematicians in centuries past—and they still impose great hardship on today's pupils.

For Pythagoras and his followers, five centuries before Christ, numbers were limited to positive integers, excluding fractions or negative numbers. Irrational quantities such as $\sqrt{2}$ were judged to be so counterintuitive that a legend says Hippasus of Metapontas was thrown overboard for proving their existence and thus shattering the Pythagorean view of a universe ruled by integers. Neither Diophantes, nor later Indian mathematicians, despite their mastery of calculation algorithms, accepted negative numbers for the solution of equations. For Pascal himself, the subtraction 0–4, whose result is negative, was pure nonsense. As for complex numbers, which were invented by Jerome Cardan in Italy in 1545 and which involve taking the square root of negative numbers their status unleashed a storm of protest that lasted over a century. We owe to Descartes, who rejected them, the epithet of "imaginary numbers," while De Morgan judged them to be "devoid of meaning, or rather self-contradictory and absurd." Only after solid mathematical foundations were established did these types of numbers gain acceptance in the mathematical community.

I would like to suggest that these mathematical entities are so difficult for us to accept and so defy intuition because they do not correspond to any preexisting category in our brain. Positive integers naturally find an echo in the innate mental representation of numerosity; hence a four-year-old can understand them. Other sorts of numbers, however, do not have any direct analogue in the brain. To really understand them, one must piece together a novel mental model that provides for intuitive understanding. This is exactly what teachers do when they introduce negative numbers with such metaphors as temperatures below zero, money borrowed from the bank, or simply a leftward extension of the number line. This is also why the English mathematician John Wallis, in 1685, made a unique gift to the mathematical community when he introduced a concrete representation of complex numbers—he first saw that they could be envisioned as a plane where the "real" numbers dwelled along a horizontal axis. To function in an intuitive mode, our brain needs images—and as far as number theory is concerned, evolution has endowed us with an intuitive picture only of positive integers.

Part II

Beyond Approximation

Chapter 4

The Language of Numbers

> I observe that when we mention any great number, such as a thousand, the mind has generally no adequate idea of it, but only a power of producing such an idea by its adequate idea of the decimals, under which the number is comprehended.
>
> David Hume, *A Treatise of Human Nature*

Suppose that our only mental representation of number were an approximate accumulator similar to the rat's. We would have rather precise notions of the numbers 1, 2, and 3. But beyond this point, the number line would vanish into a thickening fog. We could not think of number 9 without confusing it with its neighbors 8 and 10. Even if we understood that the circumference of a circle divided by its diameter is a constant, the number π would only be known to us as "about 3." This fuzziness would befuddle any attempt at a monetary system, much of scientific knowledge, indeed human society as we know it.

How did *Homo sapiens* alone ever move beyond approximation? The uniquely human ability to devise symbolic numeration systems was probably the most crucial factor. Certain structures of the human brain that are still far from understood enable us to use any arbitrary symbol, be it a spoken word, a gesture, or a shape on paper, as a vehicle for a mental representation. Linguistic symbols parse the world into discrete categories. Hence they allow us to refer to precise numbers and to separate them categorically from their closest neighbors. Without symbols, we might not discriminate 8 from 9. But with the help of our elaborate numerical notations, we can express thoughts as precise as "The speed of light is

299,792,458 meters per second." It is this transition from an approximate to a symbolic representation of numbers that I intend to describe in this chapter—a transition that occurs both in cultural history and in the mind of any child who acquires the language of numbers.

A Short History of Number

When our species first began to speak, it may have been able to name only the numbers 1, 2, and perhaps 3. Oneness, twoness, and threeness are perceptual qualities that our brain computes effortlessly, without counting. Hence giving them a name was probably no more difficult than naming any other sensory attribute, such as red, big, or warm.

The linguist James Hurford has gathered considerable evidence for the antiquity and special status of the first three number words. In languages with case and gender inflections, "one," "two," and "three" are often the only numerals that can be inflected. For instance, in old German "two" can be *zwei*, *zwo*, or *zween* depending on the grammatical gender of the object that is being counted. The first three ordinals also have a particular form. In English, for instance, most ordinals end with "th" (fourth, fifth, etc.), but the words "first," "second," and "third" do not.

The numbers 1, 2, and 3 are also the only ones that can be expressed by grammatical inflections instead of words. In many languages, words do not just bear the mark of the singular or plural. Distinct word endings are also used to distinguish two items (*dual*) versus more than two items (*plural*), and a few languages even have special inflections for expressing three items (*trial*). In ancient Greek, for instance, "o hippos" meant the horse, "to hippo" the two horses, and "toi hippoi" an unspecified number of horses. But no language ever developed special grammatical devices for numbers beyond 3.

Finally the etymology of the first three numerals also bears testimony to their antiquity. The words for "2" and "second" often convey the meaning of "another," as in the verb *to second* or the adjective *secondary*. The Indo-European root of the word "three" suggests that it might have once been the largest numeral, synonymous with "a lot" and "beyond all others"—as in the French *très* (very) or the Italian *troppo* (too much), the English *through,* or the Latin prefix *trans-*. Hence perhaps the only numbers known to Indo-Europeans were "1," "1 and another" (2), and "a lot" (3 and beyond).

Today we find it hard to imagine that our ancestors might have been confined to numbers below three. Yet this is not implausible. Up to this very day, the

Warlpiris, an aboriginal tribe from Australia, have names only for the quantities 1, 2, some, and a lot. In the domain of colors, some African languages distinguish only between black, white, and red. Needless to say, these limits are purely lexical. When Warlpiris come into contact with Occidentals, they easily learn English numerals. Thus, their ability to conceptualize numbers is not limited by the restricted lexicon of their language, nor (obviously) by their genes. Although there is a dearth of experiments on this topic, it seems likely that they possess quantitative concepts of numbers beyond three, albeit nonverbal and perhaps approximate ones.

How did human languages ever move beyond the limit of 3? The transition toward more advanced numeration systems seems to have involved the counting of body parts. All children spontaneously discover that their fingers can be put into one-to-one correspondence with any set of items. One merely has to raise one finger for the first item, two for the second, and so on. By this mechanism, the gesture of raising three fingers comes to serve as a symbol for the quantity three. An obvious advantage is that the required symbols are always "handy"—in this numeration system the digits are literally the speaker's digits!

Historically then, digits and other parts of the body have supported a body-based language of numbers, which is still in use in some isolated communities. Many aboriginal groups, who lack spoken words for numbers beyond three, possess a rich vocabulary of numerical gestures fulfilling the same role. The natives of the Torres straights, for instance, denote numbers by pointing to different parts of their body in a fixed order (Figure 4.1): from the pinkie to the thumb on the right hand (numbers 1 to 5), then up the right arm and down the left arm (6 to 12), through to the fingers of the left hand (13 to 17), the left toes (18 to 22), the left and right legs (23 to 28), and finally the right toes (29 to 33). A few decades ago, in a school in New Guinea, teachers were puzzled to see their aboriginal pupils wriggling during mathematics lessons, as if calculations made them itch. As a matter of fact, by rapidly pointing to parts of their body, the children were translating into their native body language the numbers and calculations being taught to them in English.

In more advanced numeration systems, pointing is not needed anymore: Naming a body part suffices to evoke the corresponding numeral. Thus in many societies in New Guinea, the word six is literally "wrist," while nine is "left breast." Likewise, in countless languages throughout the world, from Central Africa to Paraguay, the etymology of the word "five" evokes the word "hand."

A third step bridges the gap that separates these body-based languages from our modern "disembodied" number words. Body pointing suffers from a serious limitation: Our fingers form a finite set, indeed a rather small one. Even if we count toes and a few other salient parts of our bodies, the method is hopeless for

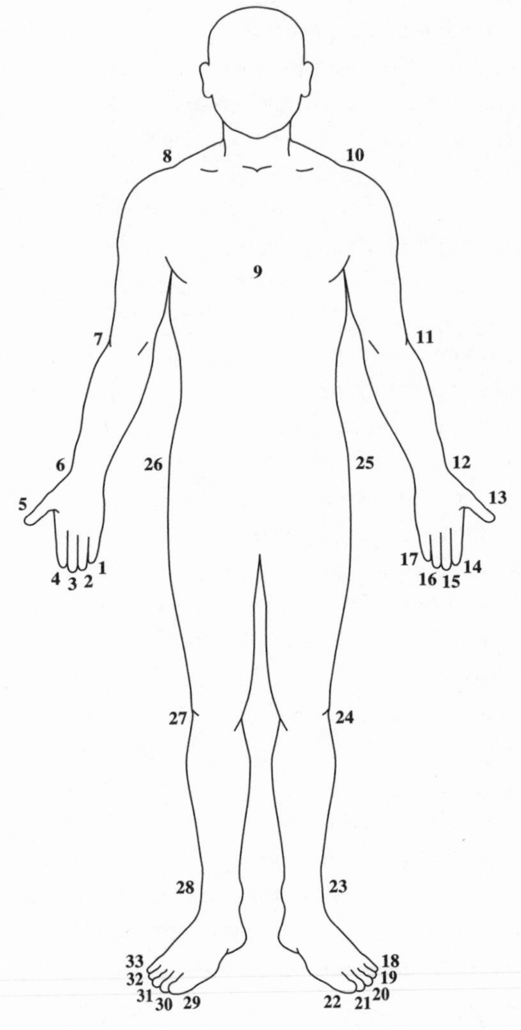

Figure 4.1. The natives of the Torres Straight denote numbers by pointing towards a precise part of their body. (After Ifrah 1994.)

numbers beyond thirty or so. It is highly impractical to learn an arbitrary name for each number. The solution is to create a syntax that allows larger numerals to be expressed by combining several smaller ones.

Number syntax probably emerged spontaneously from an extension of body-based numeration. In societies such as the native tribes of Paraguay, the number 6, instead of being given an arbitrary name such as "wrist," is expressed as "one on the other hand." Since the word "hand" itself means 5, by the very nature of their body language these people are led to express 6 as "5 and 1." Similarly, the number 7 is "5 and 2," and so on all the way to 10, which is simply expressed as "two hands" (two 5s). Behind this elementary example lurk the basic organizing principles of modern number notations: the choice of a base number (here number 5), and the expression of larger numbers by means of a combination of sums and products. Once discovered, these principles can be extended to arbitrarily large numbers. Eleven, for instance, might be expressed as "two hands and a finger" (two 5s and 1), while 22 will be "four hands and two fingers."

Most languages have adopted a base number, such as 10 or 20, whose name is often a contraction of smaller units. In the Ali language, for instance, the word "mboona," which means 10, is a contraction of "moro boona"—literally "two hands." Once the new form is frozen, it can itself enter into more complex constructions. Thus the word for 21 could be expressed as "two 10s and 1". A similar process of contraction accounts for the irregular construction of some numerals such as 11, 12, 13, or 50 in present-day English. These words were once transparent compounds—"1 (and) 10," "2 (and) 10," "3 (and) 10," "5 10s"—before they were distorted and contracted.

As for base 20, it probably reflects an ancient tradition of counting on fingers and toes. This explains why the same word often denotes number 20 and "a man," as in some Mayan dialects or in Greenland Eskimo. A number such as 93 may then be expressed by a short sentence such as "after the fourth man, 3 on the first foot"—a twisted syntax indeed, but hardly more so than the modern French expression "quatre-vingt-treize" (4×20+13). It is through such means that humans eventually learned to express any number with perfect accuracy.

Keeping a Permanent Trace of Numerals

Beyond giving numbers a name, to keep a durable record of them was also vital. For economical and scientific reasons, humans quickly developed writing systems that could maintain a permanent record of important events, dates, quantities, or exchanges—anything, in brief, that could be denoted by a number. Thus, the invention of written number notations probably unfolded in parallel with the development of oral numeration systems.

To understand the origins of number writing systems, we have to travel far back in time. Several bones from the Aurignacian period (35,000 to 20,000 B.C.) reflect the oldest method of writing numbers: the representation of a set by an identical number of notches. These bones are engraved with series of parallel notches, sometimes grouped in small packets. This might have been early humans' way of keeping a hunting record by carving one notch for each animal

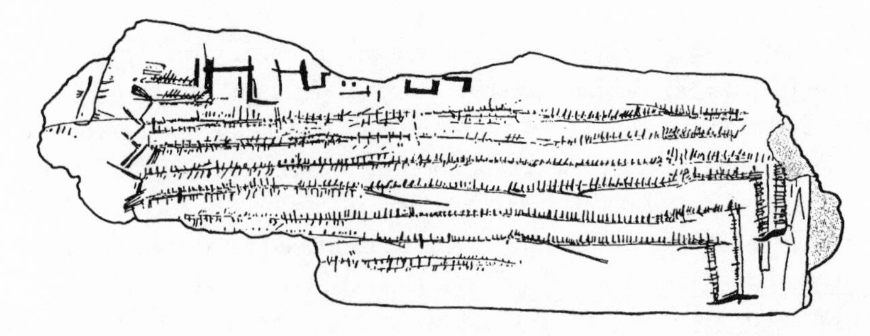

Figure 4.2. This small bone plaque was unearthed in 1969 from the Grotte du Taï in southern France. Dated from the Upper Paleolithic (ca. 10,000 BC), it is engraved with regularly aligned marks. Because some of the notches are grouped into subsets of about 29, the plaque is thought to have recorded the number of days elapsing between two lunations. (Reprinted from Marshack 1991 by permission of the publisher; copyright © 1991 by Cambridge University Press.)

captured. The patient decoding of the periodic structure of notches on a slightly more recent bone plaque even suggests that it might have been used as a lunary calendar that kept track of how many days had elapsed between two full moons (Figure 4.2).

The principle of one-to-one correspondence has been reinvented over and over again throughout the world as one of the simplest and most basic of numerical records. The Sumerians filled spheres of clay with as many marbles as the objects they counted; the Incas recorded numbers by tying knots on strings, which they kept as archives; and the Romans used vertical bars to form their first three digits. Even recently, some bakers still used notched sticks to keep track of their clients' debts. The word "calculation" itself comes from the Latin word *calculus*, which means "pebble," and draws us back to the time when numbers were manipulated by moving pebbles on an abacus.

Despite its deceptive simplicity, the one-to-one correspondence principle is a remarkable invention. It offers a durable, precise, and abstract representation of numbers. A series of notches can serve as an abstract numerical symbol and stand for any collection of items, be it livestock, people, debts, or full moons. It also enables humans to overcome the limitations of their perceptual apparatus. Humans, like pigeons, cannot distinguish forty-nine objects from fifty. Yet a stick engraved with forty-nine notches keeps a permanent track of this exact number. To verify whether a count is correct, one merely has to go through the objects one by one and move forward by one notch for each object. One-to-one correspondence therefore provides a precise representation of numbers too large to be accurately memorized on the mental number line.

Obviously, one-to-one correspondence also has its limitations. Series of notches are notoriously inconvenient to read or to write. As we have seen earlier, the human visual system is unable to apprehend at a glance the numerosity of a set of more than three items. Hence an undifferentiated series of thirty-seven notches is as difficult to perceive as the set of thirty-seven sheep it stands for! Humans were therefore quickly drawn to breaking the monotony of number series by grouping the notches and by introducing novel symbols, in effect breaking a large number into something easier to read at a glance. This is exactly what we do when we strike out each group of five strokes, thus turning them into a visually salient group. Using this technique, the number 21 looks like 𝍸 𝍸 𝍸 𝍸 |, undeniably a more readable notation than |||||||||||||||||||||.

However, this system is convenient only on paper. When engraving a stick, carving in the wood's length is tedious. Cutting the wood at an angle is so much easier, and that is exactly the method that shepherds adopted thousands of years ago: They invariably selected symbols made up of oblique bars, such as V or X, to denote the numbers 5 and 10. As you may guess, this is the origin of Roman numerals. Their geometric shapes were determined by how easily they could be carved on a wooden stick. Other writing media have imposed different shapes. For instance, the Sumerians, who wrote on sheets of soft clay, adopted for their numerals the simplest shapes that could be formed with a pencil,—namely, round or cylindrical notches as well as the famous nail-shaped or "cuneiform" characters.

By adding together several of these symbols, other numbers may be formed. In Roman notation, 7 is written as 5+1+1 (VII). This additive principle, according to which the value of a number is equal to the sum of its component digits, underlies many number notations, including those of the Egyptians, Sumerians, and Aztecs. Additive notation saves time and space, because a number such as 38, which requires thirty-eight identical symbols in any concrete notation based on one-to-one correspondence, now mobilizes only seven Roman digits (38=10+10+10+5+1+1+1 or XXXVIII). Still, reading and writing remain a tedious chore. Compactness can be improved a bit by introducing special symbols such as numbers L (50) and D (500). Repetitions may be totally avoided if one is willing to use a distinct symbol for each of the numbers 1 to 9, 10 to 90, and 100 to 900. This solution was adopted by the Greeks and the Jews, who used letters of the alphabet instead of numbers. Using this trick, a number as complex as 345 can be written with only three letters (TME in Greek, or 300+40+5). The user, however pays a heavy cost: Considerable effort is needed to memorize the numerical value of the 27 symbols required to express all numbers from 1 to 999.

In retrospect, it seems obvious that addition alone cannot suffice to express very large numbers. Multiplication becomes indispensable. One of the first hybrid notations, mixing addition and multiplication, appeared in Mesopotamia over four millennia ago. Instead of expressing a number such as 300 by repeating the symbol for 100 three times, as in Roman numerals (CCC), the inhabitants of the city of Mari simply wrote down the symbol for "three" followed by the symbol for "hundred." Unfortunately, they still wrote units and decades using the addition principle, so their notation remained far from concise. The number 2,342, for instance, was literally written down as "1+1 thousand, 1+1+1 hundred, 10+10+10+10, 1+1".

The power of the multiplication principle was refined in later number notations. In particular, five centuries ago the Chinese invented a perfectly regular notation that has been preserved up to this day. It consists of only 13 arbitrary symbols for the digits 1 through 9 and the numbers 10, 100, 1,000, and 10,000. The number 2,342 is simply written down as "2 1000 3 100 4 10 2", a word-for-word transcription of the oral expression "two thousand three hundred forty-two" (forty being "four ten" in Chinese). Thus writing, at this stage, becomes a direct reflection of the oral numeration system.

The Place-Value Principle

One final invention greatly expanded the efficacy of number notations: the place-value principle. A number notation is said to obey the place-value principle when the quantity that a digit represents varies depending on the place it occupies in the number. Thus the three digits that make up number 222, though identical, refer to different orders of magnitude: two hundreds, two tens, and two units. In a place-value notation, there is a privileged number called the base. We now use base 10, but this is not the only possibility. Successive places in the number represent successive powers of the base, from units ($10^0 = 1$), to tens ($10^1 = 10$), hundreds ($10^2 = 100$), and so on. The quantity that a given number expresses is obtained by multiplying each digit by the corresponding power of the base and then adding up all the products. Hence number 328 represents the quantity $3\times100 + 2\times10 + 8\times1$.

Place-value coding is a must if one wants to perform calculations using simple algorithms. Just try to compute XIV × VII using Roman numerals! Calculations are also inconvenient in the Greek alphabetical notation, because nothing betrays that number N (50) is ten times greater than number E (5). This is the main reason the Greeks and the Romans never performed computations without the help of an abacus. By contrast, our Arabic numerals, based on the place-value principle, make the magnitude relations between 5, 50, 500, and 5,000 completely transparent. Place-value notations are the only ones that reduce the complexity of multiplication to the mere memorization of a table of products from 2×2 up to 9×9. Their invention revolutionized the art of numerical computation.

While four civilizations seem to have discovered place-value notation, three of them never quite reached the simplicity of our current Arabic numerals. For this notation only becomes highly efficient in conjunction with three other inventions: a symbol for "zero," a unique base number, and the discarding of the addition principle for the digits 1 through 9. Consider, for instance, the oldest place-value system known, devised by Babylonian astronomers eighteen centuries before Christ. Their base number was 60. Hence a number such as 43,345, which is equal to $12 \times 60^2 + 2 \times 60 + 25$, was expressed by concatenating the symbols for 12, 2, and 25.

In principle, sixty distinct symbols would have been needed, one for each of the "digits" 0 to 59. Yet obviously it would have been impractical to learn sixty arbitrary symbols. Instead, the Babylonians wrote down these numbers using an additive base-10 notation. For instance, the "digit" 25 was expressed as $10+10+1+1+1+1+1$. Eventually, the number 43,345 was thus rendered by an obscure sequence of cuneiform characters that literally meant $10+1+1$ [implication $\times 60^2$], $1+1$ [implication $\times 60$], $10+10+1+1+1+1+1$. Such a mixture of additive and place-value coding, with two bases 10 and 60, turned the Babylonian notation into an awkward system understandable only to a cultivated elite. Still, it was a remarkably advanced numeration for its time. The Babylonian astronomers used it very skillfully for their celestial calculations, whose accuracy remained unsurpassed for more than a thousand years. Its success was due in part to its simple representation of fractions: Because 2, 3, 4, 5, and 6 are divisors of the base 60, the fractions ½, ⅓, ¼, ⅕, and ⅙ all had a simple sexagesimal expression.

Judged by today's standards, the Babylonian system had one final drawback: For fifteen centuries, it lacked a zero. What is a zero good for? It serves as a placeholder that denotes the absence of units of a given rank in a multidigit numeral. For instance, in Arabic notation the number "503" means five hundreds, no tens, and three units. Lacking a zero, Babylonian scientists simply left a blank at the place where a digit should have appeared. This meaningful void was a recurring source of ambiguities. The numbers 301 ($5 \times 60+1$), 18,001 ($5 \times 60^2+1$) and 1,080,001 ($5 \times 60^3+1$) were confusedly expressed by similar strings: 51, 5 1 (with one blank), and 5 1 (with two blanks). Hence the absence of a zero was the cause of many errors in calculation. Worse, an isolated digit such as "1" had multiple meanings. It could mean quantity 1, of course, but also "1 followed by a blank" or 1×60, or even "one followed by two blanks" or $1 \times 60^2 = 3{,}600$, and so on. Only the context could determine which interpretation was correct. Not until the third century before Christ did the Babylonians finally introduce a symbol to fill this gap and explicitly denote absent units. Even then, this symbol served only as a placeholder. It never acquired the meaning of a "null quantity" or of "the integer immediately below 1" which we attribute to it today.

While Babylonian astronomers' place-value notation was apparently lost in the collapse of their civilization, three other cultures later reinvented remarkably similar systems. Chinese scientists, in the second century before Christ, devised a place-value code devoid of the digit 0 and using the bases 5 and 10. Mayan astronomers, in the second half of the first millennium, computed with numbers written in a mixture of base 5 and 20 and with a full-fledged digit 0. And Indian mathematicians, finally, bequeathed humanity the place-value notation in base 10 that is now in use throughout the world.

It seems a bit unfair to call "Arabic numerals" an invention originally due to the ingenuity of the Indian civilization. Our number notation is called "Arabic" merely because the Western world discovered it for the first time through the mathematical writings of the great Persian mathematicians. Many of the modern techniques of numerical calculation derived from the work of Persian scientists. The word "algorithm" was named after a work by one of the them, Mohammed ibn Musa al-Khuwarizmi. His most famous book was a treatise for solving linear equations, *Al-jabr w'al muqâbala* (*On Reducing and Simplifying*), one of the few books whose publication founded a new science, "algebra." Yet for all their inventiveness, the discoveries of the Persians could not have seen the light without the help of the Indian number notation.

A particular homage should be paid to a unique innovation in the Indian notation, one that was lacking in all other place-value systems: the selection of ten arbitrary digits whose shapes are unrelated to the numerical quantities they represent. At first glance, one might think that using arbitrary shapes should be a disadvantage. A series of strokes seem to provide a more transparent way of denoting numbers, one that is easier to learn. And perhaps this was the implicit logic of the Sumerian, Chinese, and Mayan scientists. However we have seen in the preceding chapter that it is incorrect. The human brain takes longer to count five objects than to recognize an arbitrary shape and associate it with a meaning. The peculiar disposition of our perceptual apparatus for quickly retrieving the meaning of an arbitrary shape, which I have dubbed the "comprehension reflex," is admirably exploited in the Indian-Arabic place-value notation This numeration tool, with its ten easily discernible digits, tightly fits the human visual and cognitive system.

An Exuberant Diversity of Number Languages

Nowadays when people of almost any country write down a number, they adopt the same convention and employ the base-ten Arabic notation. Only the shape of digits remains slightly variable. Instead of our Arabic digits, some Middle Eastern countries such as Iran use another set of shapes referred to as "Indian digits." Even there, however, the standard Arabic notation is gaining ground. Its victory has little to do with imperialism or the establishment of commercial norms. If the evolution of written numeration converges, it is mainly because place-value coding is the best available notation. So many of its characteristics can be praised: its compactness, the few symbols it requires, the ease with which it can be learned, the speed with which it can be read or written, the simplicity of the calculation algorithms it supports. All justify its universal adoption. Indeed, it is hard to see what new invention could ever improve on it.

No such convergence is found for oral numeration. Although the vast majority of human languages possess a number syntax based on a combination of sums and products, in detail the diversity of numeration systems is striking. First of all, a variety of bases are used. In the Queensland district of Australia, some aborigines are still confined to base 2. Number 1 is "ganar," 2 is "burla," 3 "burla-ganar," and 4 "burla-burla." In old Sumer, by contrast, bases 10, 20, and 60 were concurrently used. Hence number 5,546 was expressed as "sàr (3,600) ges-u-es (60×10×3) ges-min (60×2) nismin (20×2) às (6)", or 3600 + 60×10×3+ 2×60 + 2×20 + 6 = 5546. Base 20 also had its adepts: It ruled the Aztec, Mayan, and Gaelic languages and is still in use in Eskimo and Yoruba. Traces of it can still be found in French, in which 80 is "quatre-vingt" (four twenties), and in Elizabethan English, which often counted in scores (twenty).

Although base 10 has now taken over most languages, number syntax remains highly variable. The prize for simplicity goes to Asian languages such as Chinese, whose grammar is a perfect reflection of decimal structure. In such languages there are only nine names for numbers 1 through 9 (yi, èr, san, si, wu, liù, qi, ba, and jiu), to which one should add four multipliers 10 (shi), 100 (bai), 1,000 (qian), and 10,000 (wàn). In order to name a number, one just reads its decomposition in base 10. Thus 13 is "shi san" (ten three), 27 "èr shi qi" (two ten seven), and 92,547 "shi wàn ér qian wu bai si shi qi" (nine myriads two thousands five hundreds four tens seven).

This elegant formalism contrasts sharply with the 29 words needed to express the same numbers in English or in French. In these languages, the numbers 11 through 19 and the decades from 20 to 90 are denoted by special words (eleven, twelve, twenty, thirty, etc.) whose appearance is not predictable from that of other numerals. No need to mention the even stranger peculiarities of French, with its awkward words "soixante-dix" (sixty-ten, or 70) and "quatre-vingt-dix" (four-twenty-ten, or 90). French also has confusing elision and conjunction rules involving the number 1: one says "vingt-*et*-un" (twenty-and-one) rather than "vingt-un," yet 22 is "vingt-deux" rather than "vingt-*et*-deux," and 81 is "quatre-vingt-un" and not "quatre-vingt-*et*-un." Likewise, 100 is "cent" rather than "un cent." Another eccentricity is the systematic reversal of decades and units in Germanic languages, where 432 becomes "vierhundertzweiunddreißig" (four hundred two and thirty).

What are the practical consequences of this exuberant diversity of numerical languages? Are all languages equivalent? Or are some number notations better adapted to the structure of our brains? Do certain countries, by virtue of their numeration system, start out with an advantage in mathematics? This is no trivial matter in the current period of fierce international competition, in which numeracy is a key factor to success. As adults, we are largely unaware of the complexity of our numeration system. Years of training have tamed us into accepting that 76 should be pronounced "seventy-six" rather than, say, "seven ten six" or "sixty-sixteen." Hence we can't objectively compare our language with others anymore. Rigorous psychological experiments are needed to measure the relative efficacy of various numeration systems. Surprisingly, these experiments repeatedly demonstrate the inferiority of English or French over Asian languages.

The Cost of Speaking English

Read the following list aloud: 4, 8, 5, 3, 9, 7, 6. Now close your eyes and try to memorize the numbers for twenty seconds before reciting them again. If your native language is English, you have about a 50% chance of failure. If you are Chinese, however, success is almost guaranteed. As a matter of fact, memory span in China soars to about nine digits, while it averages only seven in English. Why this discrepancy? Are speakers of Chinese more intelligent? Probably not, but their number words happen to be shorter. When we try to remember a list of digits, we generally store it using a verbal memory loop (this is why it is difficult to memorize numbers whose names sound similar, such as "five" and "nine" or "seven" and "eleven"). This memory can hold data only for about two seconds, forcing us to rehearse the words in order to refresh them. Our memory span is thus determined by how many number words we can repeat in less that two seconds. Those of us who recite faster have a better memory.

Chinese number words are remarkably brief. Most of them can be uttered in less than one-quarter of a second (for instance, 4 is "si" and 7 "qi"). Their English equivalents—"four," "seven"—are longer: pronouncing them takes about one-third of a second. The memory gap between English and Chinese apparently is entirely due to this difference in length. In languages as diverse as Welsh, Arabic, Chinese, English, and Hebrew, there is a reproducible correlation between the time required to pronounce numbers in a given language and the memory span of its speakers. In this domain, the prize for efficacy goes to the Cantonese dialect of Chinese, whose brevity grants residents of Hong Kong a rocketing memory span of about 10 digits.

In summary, the "magical number 7," which is so often heralded as a fixed parameter of human memory, is not a universal constant. It is merely the standard value for digit span in one special population of *Homo sapiens* on which more than 90% of psychological studies happen to be focused, the American college undergraduate! Digit span is a culture- and training-dependent value, and cannot be taken to index a fixed biological memory size parameter. Its variations from culture to culture suggest that Asian numerical notations, such as Chinese, are more easily memorized than our Western systems of numerals because they are more compact.

If you do not speak any Chinese, there is still hope. Several tricks are available to increase your memory for digits. First of all, always memorize numbers using the shortest possible sequence of words. A long number such as 83,412 is often best recalled by reciting it digit by digit, as with a phone number. Second, try grouping the digits into small blocks of two or three. Your working memory will jump to about twelve digits if you group them in four blocks of three. Phone numbers in the United States, with their division into a three-digit area code and then three groups of three, two, and two digits, as in "503 485 98 31," already make use of these stratagems. In France, by contrast, we have the bad habit of expressing phone numbers with two-digit numerals. For instance, we read 85 98 31 as "eighty-five ninety-eight thirty-one"—probably the most memory-inefficient method that one could think of!

A third trick is to bring the number back to familiar ground. Look for increasing or decreasing series of digits, familiar dates, zip codes, or any other information that you already know. If you can recode the number using only a few familiar items, you should easily remember them. After about 250 hours of training under the guidance of psychologists William Chase and K. Anders Ericsson, an American student was able to extend his memory span up to an extraordinary eighty digits using this recoding method. He was an excellent long-distance runner and had compiled a large mental database of record running times. He therefore stored the eighty digits to be remembered, broken down into groups of three or four, as a series of record times in long-term memory!

Using these guidelines, you should have little difficulty memorizing phone numbers. But unless you are Chinese, you are still in for a hard time. Number names also play a critical role in counting and calculating, and here again bad marks can be attributed to languages with the longest number names. For instance, it takes a Welsh pupil one second and a half more than an English pupil, on average, to compute 134+88. For equal age and education, this difference seems solely due to the time taken to pronounce the problem and the intermediate results: Welsh numerals happen to be considerably longer than the English. English is certainly not the optimum, though, because several experiments have shown that Japanese and Chinese children calculate much faster than their American peers.

It can be difficult, of course, to tease apart the effects of language from those of education, number of hours at school, parental pressure, and so on (in fact, good evidence exists that the organization of Japanese mathematics lessons is in many ways superior to that of the standard U.S. school system). However, many such variables can be left aside by studying language acquisition in children who have not yet been to school. All children are confronted with the challenging task of discovering, by themselves, the lexicon and grammar of their maternal language. How do they ever acquire the rules of French or German by mere exposure to phrases such as "soixante-quinze" or "fünfundsiebzig"? And how can a French child discover the meanings of "cent deux" and "deux cent"? Even if the child is a born linguist and if, as postulated by Noam Chomsky and Steven Pinker, the brain comes equipped with a language organ that makes learning the most abstruse linguistic rules a matter of instinct, the induction of number formation rules is by no means instantaneous and varies from language to language.

In Chinese, for instance, once you have learned the number words up to ten, the others are easily generated by a simple rule (11 = ten one, 12 = ten two . . . , 20 = two ten, 21 = two ten one, etc.). In contrast, American children have to learn by rote, not just the numerals from 1 to 10, but also those from 11 to 19, and also the tens numbers from 20 to 90. They must also discover for themselves the multiple rules of number syntax that specify, for instance, that "twenty forty" or "thirty eleven" are invalid sequences of number words.

In a fascinating experiment, Kevin Miller and his colleagues asked matched groups of American and Chinese children to recite the counting sequence. Startlingly, the linguistic difference caused American children to lag as much as one year behind their Chinese peers. When they were four, Chinese children already counted up to 40 on average. At the same age, American children painfully counted up to 15. It took them one year to catch up and reach 40 or 50. They were not just globally slower than the Chinese; up to the number 12, both groups stood on an equal footing. But when they reached the special numbers "13" and "14," American children suddenly stumbled, while the Chinese, helped by the unfailing regularity of the language, moved right along with much less trouble (Figure 4.3).

The Miller experiment shows beyond a doubt that the opacity of a numeration system takes an important toll on language acquisition. Another proof comes from the analysis of counting errors. Haven't we all heard American children recite "twenty-eight, twenty-nine, twenty-ten, twenty-eleven," and so on?

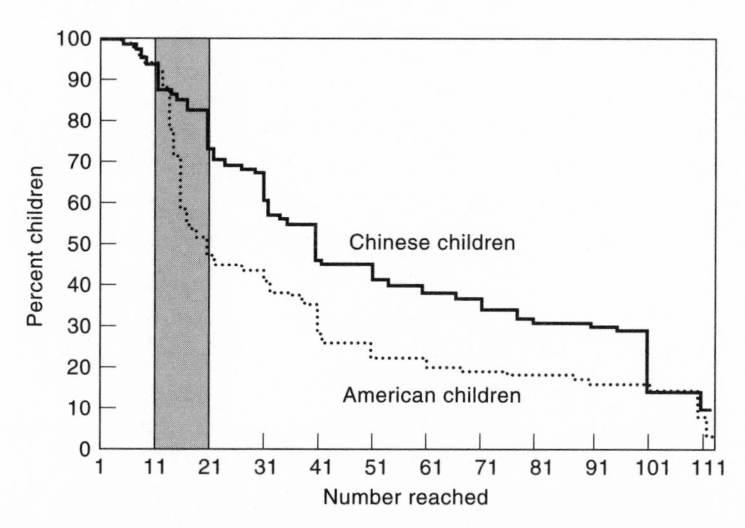

Figure 4.3. Kevin Miller and his colleagues asked American and Chinese children to recite numbers as far as they could. At a matched age, Chinese children could count much farther than their American counterparts. (Adapted from Miller et al. 1995 by permission of the publisher; copyright © 1995 by Cambridge University Press.)

Such grammatical errors, telltale signs of a poor induction of the rules of number syntax, are unheard of in Asian countries.

The influence of numeration systems carries through into subsequent school years. The organization of spoken Chinese numerals directly parallels the structure of written Arabic numerals. Hence, Chinese children experience much less difficulty than their American counterparts in learning the principles of place-value notation in base ten. When asked to form number 25 using some unit cubes and some bars of 10, Chinese schoolboys readily select two bars of 10 and five units, suggesting that they understand base ten. At a matched age, American children behave differently. Most of them laboriously count twenty-five units, thus failing to take advantage of the shortcut provided by the groups of 10. Worse yet, if one provides them with a bar comprising twenty units, they use it more frequently than two bars of ten. Thus they seem to attend to the surface form of the word "twenty-five," while the Chinese already master their deeper base-10 structure. Base 10 is a transparent concept in Asian languages, but is a real headache for Western children.

These experimental findings impose a strong conclusion: Western numeration systems are inferior to Asian languages in many respects—they are harder to keep in short-term memory, slow down calculation, and make the acquisition of counting and of base ten more difficult. Cultural selection should long have eliminated constructions as absurd as the French "quatre-vingt-dix-sept." Unfortunately, the normalization efforts of our schools and academies have put a stop to the natural evolution of languages. If children could vote, they would probably favor a widespread reform of numerical notations and the adoption of the Chinese model. Would such a revision be less utopian than the ill-fated spelling reforms? We have at least one historical example of a successful major linguistic reform. At the beginning of the twentieth century, the Welsh willingly relinquished their old numeration system, which was more complex even than present-day French, and selected instead a simplified notation quite similar to Chinese. Unfortunately, Welsh changed only to fall prey to another error: The new Welsh number words, while grammatically regular and thus easy to learn, are so long that memory suffers! Psychological experiments would probably dictate the adoption of a well-tested numeration system such as Mandarin Chinese, but national interests make this a rather distant and unlikely prospect.

Learning to Label Quantities

Acquiring a number lexicon and syntax is not everything. It is not particularly useful to know that "two hundred and thirty" is a valid English phrase while "two thirty and hundred" is not. Above all, children must learn what these numerals mean. The power of numeration systems stems from their ability to establish precise links between linguistic symbols and the quantities they express. A child may well recite numerals up to 100, but is only parroting unless he or she also knows what magnitudes they stand for. How then do children ever learn the meaning of /wan/, /siks/ or /eit/?

A first basic problem confronting a child is to recognize that these words refer to number rather than to color, size, shape, or any other dimension of the environment. Consider the phrases "the three sheep" and "the big sheep." A child who hears them for the first time and who does not know the meaning of the words "three" and "big" has no way of telling that "big" refers to the physical size of each sheep, while "three" refers to the cardinal of the set of sheep.

Experiments show that by two and a half years of age, American children already differentiate number words from other adjectives. When given a choice between a picture of a single red sheep and another showing three blue sheep, children readily point to the first when they are told, "Show me the red sheep," and to the second when told, "Show me the three sheep." By that age, children already know that "three" applies to a collection of items rather than to a single item. At the same age, children also order number words and other adjectives correctly. They say "three little sheep," but never "little three sheep." Early on, then, children know that number words belong to a special category distinct from other words.

How did they find this out? Probably by exploiting all the available cues, be they grammatical or semantic. Grammar alone may be of precious help. Suppose that a mother tells her baby, "Look, Charlie, three little doggies." Baby Charlie may then infer that the word "three" is a special kind of adjective because other adjectives such as "nice," are always said with an article—"the nice little doggies." The fact that the word "three" does not require an article may suggest that "three" applies to the entire collection of little doggies, and that therefore it may be a number or a quantifier like "some" or "many."

Of course such reasoning is of little help for determining the precise quantity to which the word "three" refers. Indeed, it appears that for a whole year, children realize that the word "three" is a number without knowing the precise value it refers to. When they are ordered, "Give me three toys," most of them simply grasp a pile without caring about the exact number. If one lets them choose between a group of two and a group of three toys, they also respond at random—although they never select a card showing a single object. They know how to recite number words, and they sense that these words have to do with quantity, but they ignore their exact meaning.

Semantic cues are probably critical in order to overcome this stage and to determine the precise quantity that is meant by the word "three." With a little luck, Baby Charlie will see the three little doggies his mom is talking about. His perceptual system, whose sophistication we have discussed in Chapter 2, may then analyze the scene and identify the presence of several animals, of a small size, noisy, moving, and numbering about three. (By this I do not mean, of course, that Charlie already knows that the word "three" applies to this numerosity; I only mean that Charlie's internal non-verbal accumulator has reached the state of fullness which is typical of sets of three items.)

In essence, all Charlie has to do, then, is to correlate these preverbal representations with the words he hears. After a few weeks or months, he should realize that the word "three" is not always uttered in the presence of small things, of animals, of movement, or of noise; but that it is very often mentioned when his mental accumulator is in a particular state that accompanies the presence of three items. Thus, correlations between number words and his prior nonverbal numerical representations can help him determine that "three" means 3.

This correlation process can be accelerated by the "principle of contrast," which stipulates that words that sound different have different meanings. If Charlie already knows the meaning of the words "doggie" and "small," the principle of contrast guarantees him that the unknown word "three" cannot refer to the size or the identity of the animals. Narrowing down the set of hypotheses enables him to find out even faster that this word refers to numerosity three.

Round Numbers, Sharp Numbers

Once children have acquired the exact meaning of number words, they still have to grasp some of the conventions governing their use in language. One of them is the distinction between round numbers and sharp numbers. Let me introduce it with a joke:

> At the museum of natural history, a visitor asks the curator, "How old is this dinosaur over here?" "Seventy million and thirty-seven years" is the answer. As the visitor marvels at the accuracy of the dating, the curator explains: "I've been working here for 37 years, you know, and when I arrived I was told that it was 70 million years old"!

Lewis Carroll, well-known for his ingenious word games based on logic and mathematics, often spiced his stories with "numerical non-sense." Here is an example from his little-known book *Sylvie and Bruno Concluded*:

> "Don't interrupt," Bruno said as we came in. "I'm counting the Pigs in the field!"
> "How many are there?" I enquired.
> "About a thousand and four," said Bruno.
> "You mean 'about a thousand,'" Sylvie corrected him. "There's no good saying 'and four': you ca'n't be sure about the four!"
> "And you're as wrong as ever!" Bruno exclaimed triumphantly. "It's just the four I can be sure about; 'cause they're here, grubbling under the window! It is the thousand I isn't pruffickly sure about!"

Why do these exchanges sound eccentric? Because they violate an implicit and universal principle governing the use of numerals. The principle stipulates that certain numerals, called "round numbers," can refer to an approximate quantity while all other numerals necessarily have a sharp and precise meaning. When one states that a dinosaur is 70 million years old, this value is implicitly understood to within 10 million years. The rule is that a number's accuracy is given by its last non-zero digit starting from the right. If I maintain that the population of Mexico City is 39,000,000, I mean that this number is correct to within a million, whereas if I give you a value of 39,452,000 inhabitants, I implicitly admit that it is correct to give or take a thousand.

This convention sometimes leads to paradoxical situations. If a precise quantity happens to fall exactly on a round number, just asserting it is not sufficient. One must supplement it with an adverb or locution that makes its accuracy explicit—for example, "Today, Mexico has *exactly* 39 million inhabitants." For the same reason, the sentence "nineteen is about 20" is acceptable, while "twenty is about 19" isn't. The phrase "about 19" is a contradiction in terms, for why use a sharp number such as 19 if one wants to state an estimation?

All the languages of the world seem to have selected a set of round numbers. Why this universality? Probably because all humans share the same mental apparatus and are therefore all confronted with the difficulty of conceptualizing large quantities. The larger a number, the less accurate our mental representation of it. Language, if it wants to be a faithful vehicle for thought, must incorporate devices that express this increasing uncertainty. Round numbers are such a device. Conventionally, they refer to approximate quantities. The sentence "There are twenty students in this room" remains true even if there are eighteen or twenty-two students because the word "twenty" can refer to an extended region of the number line. This is also why speakers of French find it so natural that "fifteen days" means "two weeks," although the exact number should be fourteen.

Approximation is so important to our mental life that many other linguistic mechanisms are available to express it. All languages possess a rich vocabulary of words for expressing various degrees of numerical uncertainty—about, around, circa, almost, roughly, approximately, more or less, nearly, barely, and so on. Most languages have also adopted an interesting construction in which two juxtaposed numbers, often linked by the conjunction "or," express a confidence interval: two or three books, five or ten people, a boy age twelve or fifteen years, 300 or 350 dollars. This construction allows us to communicate not just an approximate quantity, but also the degree of accuracy that should be granted to it. Thus the same central tendency can be expressed with increasing uncertainty by saying 10 or 11, 10 or 12, 10 or 15, or 10 or 20.

A linguistic analysis by Thijs Pollmann and Carel Jansen shows that two-number constructions follow certain implicit rules. Not all intervals are equally acceptable. At least one of the numbers must be round: One can say "twenty or twenty-five dollars" but not "twenty-one or twenty-six dollars." The other number must be of a similar order of magnitude: "Ten or one thousand dollars" sounds very strange indeed. Another Lewis Carroll quote illustrates this point:

> "How far have you come, dear?" the young lady persisted.
> Sylvie looked puzzled. "A mile or two, I think," she said doubtfully.
> "A mile or three," said Bruno.
> "You should not say 'a mile or three,'" Sylvie corrected him.

The young lady nodded approval. "Sylvie's quite right. It isn't usual to say 'a mile or three.'"

"It would be usual—if we said it often enough," said Bruno.

Bruno is wrong—"a mile or three" would never sound right, because it violates the basic rules of the two-number construction. These rules are understandable if one considers which representations we intend to communicate. These representations are fuzzy intervals on a mental number line. When we say "twenty, twenty-five dollars," we actually mean "a certain fuzzy state of my mental accumulator, somewhere around 20 and with a variance of about 5." Neither the interval from 21 to 26, nor that from 10 to 1,000, or from 1 to 3 are plausible states of the accumulator because the former is too accurate while the latter two are too imprecise.

Why Are Some Numerals More Frequent Than Others?

Would you like to try a bet? Open a book at random and note the first digit that you encounter. If this digit is either 4, 5, 6, 7, 8, or 9, you win ten dollars. If it is 1, 2, or 3, I win this amount. Most people are ready to take this bet, because they believe that the odds are 6:3 for them to win. And yet the bet is a loser. Believe it or not, the digits 1, 2, and 3 are about twice as likely to appear in print than all other digits combined!

This is a strongly counterintuitive finding because the nine digits seem equivalent and interchangeable. But we forget that numbers that appear in print are not drawn from a random number generator. Each of them represents an attempt to transmit a piece of numerical information from one human brain to another. Hence, how frequently each numeral is used is determined in part by how easily our brain can represent the corresponding quantity. The decreasing precision with which numbers are mentally represented influences not just the perception, but also the production of numerals.

Jacques Mehler and I have systematically looked for number words in tables of word frequency. Such tables tally up how often a certain word, say "five," appears in written or spoken texts. Frequency tables are available in a great variety of languages, from French to Japanese, English, Dutch, Catalan, Spanish, and even Kannada, a Dravidian language spoken in Sri Lanka and southern India. In all of these languages, despite enormous cultural, linguistic, and geographic diversity, we have observed the same results: The frequency of numerals decreases systematically with number size.

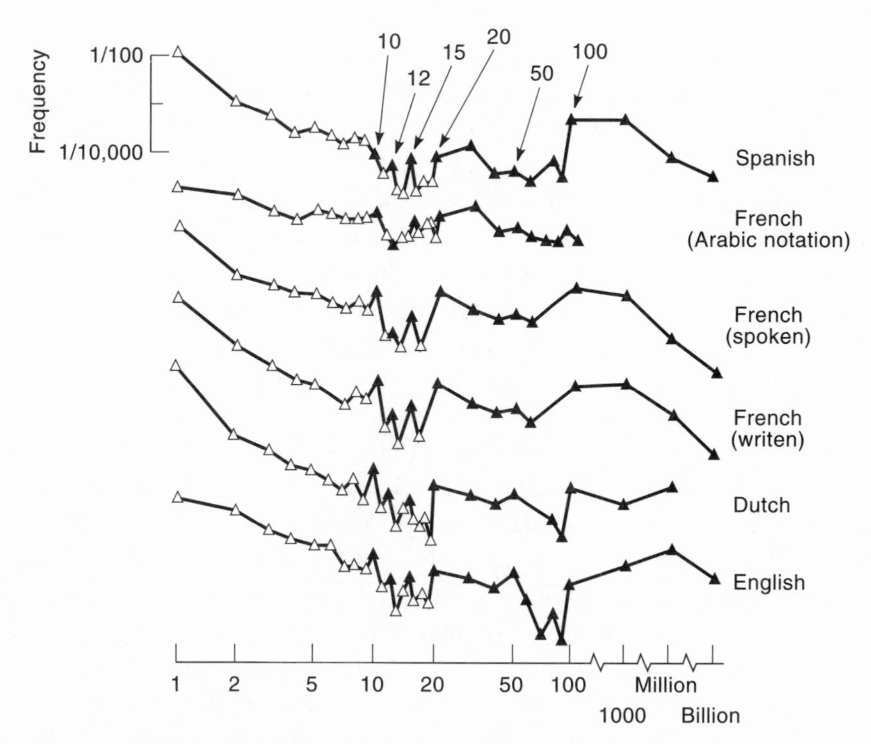

Figure 4.4. In all languages, the frequency with which number words are printed or uttered decreases with magnitude, aside from local increases for the round numbers 10, 12, 15, 20, 50, and 100. For instance, we read or hear the word two *about ten times more often than the word* nine. *(After Dehaene and Mehler 1992.)*

In French, for instance, the word "un" appears once every seventy words or so, the word "deux" once every six hundred words, the word "trois" once every one thousand, seven hundred words, and so on. Frequency decreases from 1 to 9, but also from 11 to 19, and for tens numbers from 10 to 90. A similar decrease is observed for written or spoken numerals, for Arabic numerals, and even for ordinals from "first" to "ninth." It is accompanied by a few deviations that are also universal: the very low frequency of the word "zero," and the elevated peaks for 10, 12, 15, 20, 50, and 100 (Figure 4.4). Remarkably, such cross-linguistic regularities persist in the face of pronounced differences in the way numbers are expressed, such as the absence of teen words in Japanese, the inversion of tens and units in Dutch, or the cryptic base 20 of the French words 70, 80, and 90.

I contend that once again, these linguistic regularities reflect the way our brain represents numerical quantities. Yet before jumping to this conclusion, several alternative explanations have to be examined. Ambiguity may be a possible source of this finding. In many languages the word for "one" is indistinguishable from the indefinite article "a." This probably contributes to the elevated frequency of the word "un" in French—but obviously not in English, where "one" can only be a number word. Ambiguity is also not a problem beyond "two," and yet frequency decreases sharply beyond this point.

Another contributing factor is our propensity for counting, which implies that many objects in our environment are numbered starting at 1. In any city, more houses bear number 1 than number 100, merely because all streets have a number 1, but some don't reach number 100. Again this effect certainly contributes to the elevated frequency of small numerals, but quick calculation shows that by itself it cannot account for the exponential drop of number frequency, even in the interval from 1 to 9.

Purely mathematical explanations of the effect should also be given some consideration. Few people know the following *very* counterintuitive mathematical law: If you draw several random numbers from essentially any smooth distribution, the numbers will start more often with 1 than with 9. This singular phenomenon is called Benford's law. Frank Benford, an American physicist, made a curious observation: At his university's library, the first pages of the tables of logarithms were more worn out than the last. Now surely people did not read tables of logarithms like a bad novel, stopping halfway through. Why did his colleagues have to consult the beginning of the table more often than the end? Could it be that small numbers were used more often than large ones? To his own bewilderment, Benford discovered that numbers of all origins—the surface of American lakes, his colleagues' street addresses, the square root of integers, and so on—were about six times more likely to start with digit 1 than with digit 9. About 31% of numbers started with 1, 19% with 2, 12% with 3, and the percentages decreases with each successive number. The probability that a number started with digit n was very accurately predicted by the formula $P(n) = \log_{10}(n+1) - \log_{10}(n)$.

The exact origin of this law is still poorly understood, but one thing is certain: This is a purely formal law, due solely to the grammatical structure of our numerical notations. It has nothing to do with psychology: A computer reproduces it when it prints random numbers in Arabic notation or even spells them out. The only constraint seems to be that the numbers be drawn from a sufficiently smooth distribution spread over several orders of magnitude—for instance, from 1 to 10,000.

Benford's law certainly contributes to amplifying the frequency of small numbers in natural language. Yet its explanatory power is limited. The law applies only to the frequency of the leftmost digit in a multidigit numeral, and so it does not have any influence on how frequently we refer to the quantities 1 through 9.

But the measurements that Jacques Mehler and I have performed show, quite straightforwardly, that the human brain finds it more important to talk about quantity 1 than about quantity 9. Contrary to Benford's law, this fact has nothing to do with the production of large multidigit numerals.

If it is not the grammar of numerical notations that drives us to produce small numbers, could it be Mother Nature herself? Aren't small collections of objects exceptionally frequent in our environment? To take just one example, discussing the number of one's children nowadays usually only requires number words below 3 or 4! Yet as a general explanation of the decreasing frequency of numerals, this account is misguided. Philosophers Gottlob Frege and W.V.O. Quine have long demonstrated that, objectively speaking, small numerosities are no more frequent than larger ones in our environment. In any situation, a potential infinity of things might be enumerated. Why do we prefer to speak of *one* deck of cards rather than fifty-two cards? The notion that the world is mostly made up of small sets is an illusion imposed on us by our perceptual and cognitive systems. Nature isn't made that way, no matter what our brain may think.

To prove this point without resorting to philosophical arguments, consider the distribution of words with a numerical prefix, such as "*bi*cycle" or "*tri*angle." Just as the word "two" is more frequent than "three," there are more words that begin with the prefix *bi* (or *di* or *duo*) that with *tri*. Crucially, this remains true even in domains where there is arguably little or no environmental bias for small numbers. Consider time. My English dictionary lists fourteen temporal words with the prefix *bi* or *di* (from "biannual" to "diestrual"), five words with the prefix *tri* (from "triennial" to "triweekly"), five words with a prefix expressing fourness, and only two expressing fiveness (the uncommon words "quinquennial" and "quinquennium"). Hence, increasingly fewer words express increasingly large numbers. Could this be due to an environmental bias? In the natural world, events do not recur particularly often with a two-month period. No, the culprit is our brain, which pays more attention to events when they concern small or round numbers.

If a lexical bias for small numbers can emerge in the absence of any environmental bias, conversely there are situations in which an objective bias fails to be incorporated in the lexicon. Many more vehicles have four wheels than two, yet we have a number-prefixed word for the latter (bicycle) but not for the former (quadricycle?). Numerical regularities in the world seem to be lexicalized only if they concern a small enough numerosity. For instance, we have number-prefixed words for plants with three leaves (trifoliate, trifolium; trèfle in French), but not for the many other plants or flowers with a fixed but large number of leaves or petals. Words like "octopus" that explicitly refer to a precise large numerosity are rare. As a final example, *Scolopendra morsitans*, an arthropod with twenty-one body segments and forty-two legs, is commonly called a centipede (one hundred feet) in English and a "mille-pattes" (thousand-legs) in French! Clearly, we pay attention to the numerical regularities of nature only inasmuch as they fit in with our cognitive apparatus, which is biased toward small or round numerosities.

Human language is deeply influenced by a nonverbal representation of numbers that we share with animals and infants. I believe that this alone explains the universal decrease of word frequency with number size. We express small numbers much more often than large ones because our mental number line represents numbers with decreasing accuracy. The larger a quantity is, the fuzzier our mental representation of it, and the less often we feel the need to express that precise quantity.

Round numbers are exceptions because they can refer to an entire range of magnitudes. This is why the frequency of the words "ten," "twelve," "fifteen," "twenty," and "hundred" is elevated compared to their neighbors. All in all, both the global decrease and the local peaks in number frequency can be explained by a labeling of the internal number line (Figure 4.5). As children acquire language, they learn to put a name on each range of magnitudes. They discover that the word "two" applies to a percept that they know from birth; that "nine" pertains only to the precise quantity 9, which is difficult to represent exactly; and that people often use the word "ten" to mean any quantity somewhere between 5 and 15. In turn, they therefore utter the words "two" and "ten" more often than "nine," hence perpetuating the lawful distribution of number frequencies.

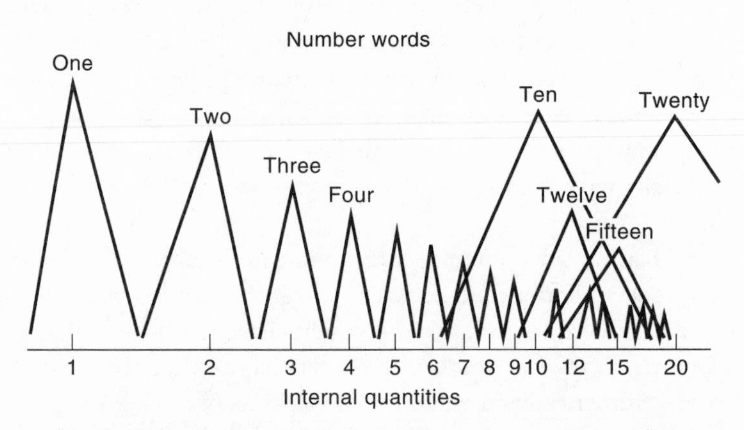

Figure 4.5. The decreasing frequency of numerals is due to the organization of our mental representation of quantities. The larger a number, the less accurate our mental representation of it; hence, the less often we need to use the corresponding word. As for round numbers like 10, 12, 15 or 20, they are uttered more frequently than others because they can refer to a greater range of quantities. (After Dehaene and Mehler 1992.)

One last detail: Our study showed that, in all Western languages, the frequency of number 13 was lower than that of 12 or 14. This seems the result of the "Devil's dozen" superstition, which assigns a maleficent power to number 13 and is known to such a degree that many American skyscrapers have no 13th floor. In India, where this superstition is unknown, the frequency of the number 13 does not show any notable drop. The frequency of numerals appears to faithfully reflect their importance in our mental lives, even in their most trivial details.

Cerebral Constraints on Cultural Evolution

What does the analysis of numerical languages reveal about the relationship between mathematics and the brain? It shows that numeration systems have evolved both *through* the brain and *for* the brain. Through the brain, because the history of number notations is clearly limited by the inventiveness of the human brain and its ability to fathom new principles of numeration. For the brain, because numerical inventions have been transmitted from generation to generation only when they closely matched the limits of human perception and memory, and therefore increased humankind's computational potential.

The history of numerals is obviously not driven merely by random factors. It exhibits discernible regularities that transcend the fortunes of history. Across borders and oceans, men and women of all colors, cultures, and religions have regularly reinvented the same notation devices. The place-value principle was rediscovered, with an interval of about three thousand years, in the Middle East, on the American continent, in China, and in India. In all languages, frequency decreases with number size. In all languages too, round numbers are contrasted with sharp numbers. The explanation of these striking cross-cultural parallels does not reside in dubious exchanges between remote civilizations. Rather, they discovered similar solutions because they were confronted with the same problems and have been endowed with the same brain for solving them.

Let me sketch a summary of humankind's slow march toward greater numerical efficacy—a summary that must remain highly schematic, given that history is rarely linear and that some cultures may have skipped several steps.

1. Evolution of oral numeration

Starting point: The mental representation of numerical quantities that we share with animals

PROBLEM: How to communicate these quantities through spoken language?
SOLUTION: Let the words "one," "two," and "three" refer directly to the subitized numerosities 1, 2, and 3.

PROBLEM: How to refer to numbers beyond 3?
SOLUTION: Impose a one-to-one correspondence with body parts (12 = pointing to the left breast).

PROBLEM: How to count when the hands are busy?
SOLUTION: Turn the names of body parts into number names (12 = "left breast").

PROBLEM: There is only a limited set of body parts, compared with an infinity of numbers.
SOLUTION: Invent number syntax (12 = "two hands and two fingers").

PROBLEM: How to refer to approximate quantities?
SOLUTION: Select a set of "round numbers" and invent the two-word construction (e.g., ten or twelve people).

2. Evolution of written numeration

PROBLEM: How to keep a permanent trace of numerosities?
SOLUTION: One-to-one correspondence. Engrave notches on bone, wood, and so on (7 = IIIIIII).

PROBLEM: This representation is hard to read.
SOLUTION: Regroup the notches (7 = IIIII II). Replace some of these groups with a single symbol (7 = VII).

PROBLEM: Large numbers still require many symbols (e.g., 37 = XXXVII).
IMPASSE 1: Add even more symbols (e.g., L instead of XXXXX).
IMPASSE 2: Use distinct symbols to denote units, tens, and hundreds (345 = TME).
SOLUTION: Denote numbers using a combination of multiplication and addition (345 = 3 hundreds, 4 tens, and 5).
PROBLEM: This notation still suffers from the repetition of the words "hundreds" and "tens."
SOLUTION: Drop these words, resulting in a shorter notation ancestral to modern place-value notation (437 = 4 3 7).

PROBLEM: This notation is ambiguous when units of a certain rank are lacking (407, denoted as 4 7, is easily confused with 47).
SOLUTION: Invent a placeholder, the symbol zero.

The cultural evolution of numeration systems testifies to the inventiveness of humanity. Across centuries, ingenious notation devices have been invented and constantly refined, the better to fit the human mind and improve the usability of numbers. The history of number notations is hard to reconcile with the Platonist conception of numbers as ideal concepts that transcend humankind and give us access to mathematical truths independent of the human mind. Contrary to what the Platonist mathematician Alain Connes has written, mathematical objects are not "untainted by cultural associations"—or at the very least this is not true of numbers, one of the most central of all mathematical objects. What has driven the evolution of numeration systems is obviously not an "abstract concept" of number, nor an ethereal conception of mathematics. If this were the case, as generations of mathematicians have noted, binary notation would have been a much more rational choice than our good old base 10. At least a prime number such as 7 or 11, or perhaps a number with many divisors such as 12, should have been selected as the base of numeration. But more prosaic criteria governed our ancestors' choices. The preponderance of base 10 is due to the contingent fact that we have ten fingers; the bounds of our subitization procedure account for the structure of Roman numerals; and the sharp limits of our short-term memory explain the constant drive toward a compact notation for large numbers. Let us leave the last word to the philosopher Karl Popper: "The natural numbers are the work of men, the product of human language and of human thought."

Chapter 5

Small Heads for Big Calculations

Two and two make four
four and four make eight
eight and eight make sixteen
Repeat!, says the teacher
Jacques Prévert, *Page d'écriture*

Ambition, distraction, uglification, and derision. These are the mischievous names the Reverend Charles Lutwidge Dodgson, a mathematics professor better known to us as Lewis Carroll, gave to the four arithmetical operations. Obviously, Carroll did not cherish too many illusions about his pupils' calculation abilities. And perhaps he was right. While children easily acquire number syntax, learning to calculate can be an ordeal. Children and even adults often err in the most elementary of calculations. Who can say that they never get 7×9 or 8×7 wrong? How many of us can mentally compute 113–37 or 100–24 in less than two seconds? Calculation errors are so widespread that far from stigmatizing ignorance, they attract sympathy when they are admitted publicly ("I've always been *hopeless* at math!"). Many of us can almost identify with Alice's plight as she attempts to calculate while traveling through Wonderland: "Let me see: four times five is twelve, and four times six is thirteen, and four times seven is—oh dear! I shall never get to twenty at that rate!"

Why is mental calculation so difficult? In this chapter, we examine the calculation algorithms of the human brain. Although our knowledge of this issue is still far from complete, one thing is certain: Mental arithmetic poses serious problems for the human brain. Nothing ever prepared it for the task of memorizing dozens of intermingled multiplication facts, or of flawlessly executing the ten or fifteen steps of a two-digit subtraction. An innate sense of approximate numerical quantities may well be embedded in our genes; but when faced with exact symbolic calculation, we lack proper resources. Our brain has to tinker with alternate circuits in order to make up for the lack of a cerebral organ specifically designed for calculation. This tinkering takes a heavy toll. Loss of speed, increased concentration, and frequent errors illuminate the shakiness of the mechanisms that our brain contrives in order to "incorporate" arithmetic.

Counting: The ABC of Calculation

In the first six or seven years of life, a profusion of calculation algorithms see the light. Young children reinvent arithmetic. Spontaneously or by imitating their peers, they imagine new strategies for calculation. They also learn to select the best strategy for each problem. The majority of their strategies are based on counting, with or without words, with or without fingers. Children often discover them by themselves, even before they are taught to calculate.

Does this imply that counting is an innate competence of the human brain? Rochel Gelman and Randy Gallistel, from the psychology department of UCLA, have championed this point of view. According to them, children are endowed with unlearned principles of counting. They do not have to be taught that each object must be counted once and only once, that the number words must be recited in fixed order, or that the last number represents the cardinal of the whole set. Gelman and Gallistel maintain that such counting knowledge is innate and even precedes and guides the acquisition of the number lexicon.

Few theories have been as harshly debated as that of Gelman and Gallistel's. For many psychologists and educators, counting is a typical example of learning by imitation. Initially, it is just a rote behavior devoid of meaning. According to Karen Fuson, children initially recite "onetwothreefourfive . . . " as an uninterrupted chain. Only later do they learn to segment this sequence into words, to extend it to larger numerals, and to apply it to concrete situations. They progressively infer what counting is about by observing other people count. Initially, according to Fuson, counting is just parroting.

The truth, which is being progressively unveiled after years of controversy and tens of experiments, seems to stand somewhere between the "all innate" and the "all acquired" extremes. Some aspects of counting are mastered quite precociously, while others seem to be acquired by learning and imitation.

As an example of an amazingly precocious competence for counting, consider the following experiment by Karen Wynn. At two and a half, children have probably not had many occasions to see someone count sounds or actions. Yet if one asks them to watch a *Sesame Street* videotape and count how many times Big Bird jumps, they easily lend themselves to this task. Likewise, they can count sounds as diverse as trumpeting, a bell, a splash, and a computer beep that have been recorded on tape and whose source is not visible. So children seem to understand, quite early on and without explicit teaching, that counting is an abstract procedure that applies to all kinds of visual and auditory objects.

Here is another precocious competence: As early as three and half years of age, children know that the order in which one recites numerals is crucial, while the order in which one points toward objects is irrelevant as long as each object is counted once and only once. In an innovative series of experiments, Gelman and her colleagues presented children with situations that violate the usual conventions of counting. The results indicate that three-and-a-half-year-olds can identify and correct rather subtle counting errors. They never fail to notice when someone recites numerals out of order, forgets to count an item, or counts the same item twice. Most important, they maintain a clear distinction between such patent errors and other correct though unusual ways of counting. For instance, they find it perfectly acceptable to start counting at the middle of a row of objects, or to count every other object first, as long as one eventually counts all items once and only once. Better yet, they are willing to start counting at any point in a row, and they can even devise strategies to systematically reach a predesignated object in third position.

What these experiments show is that by their fourth year, children have mastered the basics of how to count. They are not content with slavishly imitating the behavior of others: They generalize counting to novel situations. The origins of this precocious competence remain poorly understood. From where does a child draw the idea of reciting words in a perfect one-to-one correspondence with the objects to be counted? Like Gelman and Gallistel, I believe that this aptitude belongs to the genetic endowment of the human species. Reciting words in a fixed order is probably a natural outcome of the human faculty for language. As to the principle of one-to-one correspondence, it is actually widespread in the animal kingdom. When a rat forages through a maze, it tries to visit each arm once and only once, a rational behavior that minimizes exploration time. When we look for a given object in a visual scene, our attention is oriented in turn toward each object. The counting algorithm stands at the intersection of these two elementary abilities of the human brain—word recitation and exhaustive search. That is why our children easily dominate it.

Though children rapidly grasp the *how to* of counting, however, they seem to initially ignore the *why*. As adults, we know what counting is for. To us, counting is a tool that serves a precise goal: enumerating a set of items. We also know that what really matters is the final numeral, which represents the cardinal of the entire set. Do young children have this knowledge? Or do they just view counting as an entertaining game in which one recites funny words while pointing to various objects in turn?

According to Karen Wynn, children do not appreciate the meaning of counting until the end of their fourth year. Let your three-year-old daughter count up her toys and then ask her, "How many toys do you have?" Chances are, she will give a random number, not necessarily the one she just reached. Like all children of this age, she does not seem to relate the "how many" question to her previous counting. She may even count everything up again, as if the act of counting itself was an adequate answer to a "how many" question. Likewise, ask a two-and-a-half-year-old boy to give you three toys. Most likely he will pick a handful at random, even if he can already count up to five or ten. At that age, although the mechanisms of counting have largely fallen into place, children do not seem to understand what counting is for, and they do not think of counting when the situation commands it.

Around four, the meaning of counting eventually settles in. But how? The preverbal representation of numerical quantities probably plays a crucial role in this process. Remember that right from birth, way before they start to count, children have an internal accumulator that informs them of the approximate number of things that surround them. This accumulator can help bring meaning to counting. Suppose that a child is playing with two dolls. His accumulator automatically activates a cerebral representation of the quantity 2. Thanks to the processes described in an earlier chapter, the child has learned that the word "two" applies to this quantity, so that he can say "two dolls" without having to count. Now suppose that for no particular reason, he decides to "play the counting game" with the dolls, and recites the words "one, two." He will be surprised to discover that the last number of the count, "two," is the very word that can apply to the entire set. After ten or twenty such occasions, he may soundly infer that whenever one counts, the last word arrived at has a special status: It represents a numerical quantity that matches the one provided by the internal accumulator. Counting, which was only an entertaining word game, suddenly acquires a special meaning: Counting is the best way of *saying how many*!

Preschoolers as Algorithm Designers

Understanding what counting is for is the starting point of an outburst of numerical inventions. Counting is the Swiss Army knife of arithmetic, the tool that children spontaneously put to all sorts of uses. With the help of counting, most children find ways of adding and subtracting numbers without requiring any explicit teaching.

The first calculation algorithm that all children figure out for themselves consists in adding two sets by counting them both on the fingers. Ask a very young child to add 2 and 4. She will typically start by counting up to the first number, 2, while successively raising two fingers. Then she will count up to the second number, 4, while raising four other fingers. And finally she will recount them all and reach a total of 6. This first "digital" algorithm is conceptually simple but very slow. Executing it can be truly awkward: At the age of four, to compute 3+4, my son would put up three fingers on the left hand and four on the right hand. Then he would proceed to count them using the only pointing device that remained at his disposal—the tip of his nose!

Initially, young children find it difficult to calculate without using their fingers. Words vanish as soon as they have been uttered, but fingers can be kept constantly in sight, preventing one from losing count in case of a temporary distraction. After a few months though, children discover a more efficient addition algorithm than finger counting. When adding two and four, they can be heard muttering "one *two* . . . three . . . four . . . five . . . *six*." They first count up to the first operand, 2, then move forward by as many steps as specified by the second operand, 4. This is an attention-demanding strategy because it implies some sort of recursion: In the second phase, one has to count how many times one counts! Children often make this recursion explicit: "one *two* . . . three is one . . . four is two . . . five is three . . . six is four . . . *six*." The difficulty of this step is reflected by a drastic slowing and extreme concentration.

Refinements are quickly found. Most children realize that they need not recount both numbers, and that they can compute 2+4 by starting right from the word "two." They then simply say "two . . . three . . . four . . . five . . . *six*." To shorten calculation even further, they learn to systematically start with the larger of the two numbers. When asked to compute 2+4, they spontaneously transform this problem into the equivalent 4+2. As a result, all they now have to do is count a number of times equal to the smaller of the two addends. This is called the "minimum strategy." It is a standard algorithm that underlies most of children's calculations before the onset of formal schooling.

It is rather remarkable that children spontaneously think of counting from the larger of the two numbers to be added. This indicates that they have a very precocious understanding of the commutativity of addition (the rule that $a+b$ is always equal to $b+a$). Experiments show that this principle is already in place by five years of age. Never mind the legions of educators and theorists who have claimed that children couldn't possibly understand arithmetic unless they first received years of solid education in logic. The truth is just the opposite: As children count on their fingers, years before going to school, they develop an intuitive understanding of commutativity, whose logical foundations they will come to appreciate only much later (if ever).

Children select their calculation algorithms with an extraordinary flair. They quickly master many addition and subtraction strategies. Yet far from being lost in this abundance of possibilities, they learn to carefully select the strategy that seems most suited for each particular problem. For 4+2, they may decide to count on from the first operand. For 2+4, they will not forget to reverse the two operands. Confronted with the more difficult 8+4, they might remember that 8+2 is 10. If they manage to decompose 4 into 2+2, then they'll be able to simply count "ten, eleven, *twelve.*"

Calculation abilities do not emerge in an immutable order. Each child behaves like a cook's apprentice who tries a random recipe, evaluates the quality of the result, and decides whether or not to proceed in this direction. Children's internal evaluation of their algorithms takes into account both the time it takes them to complete the computation and the likelihood that they have reached the correct result. According to child psychologist Robert Siegler, children compile detailed statistics on their success rate with each algorithm. Little by little, they acquire a refined database of the strategies that are most appropriate for each numerical problem. There is no doubt that mathematical education plays an extremely important role in this process, both by inculcating new algorithms into children and by providing them with explicit rules for selecting the best strategy. Yet the best part of this process of invention followed by selection is established in most children before they even reach their preschool years.

Would you like a final example of children's shrewdness in designing their own calculation algorithms? Consider the case of subtraction. Ask a young boy to compute 8–2, and you may hear him muttering: "eight . . . seven is one . . . six is two . . . *six*": He counts backward starting from the larger number 8. Now ask him to solve 8–6. Does the child have to count backward "eight seven six five four three two"? No. Chances are, he will find a more expeditious solution: "six . . . seven is one . . . eight is two . . . *two*!" He counts the number of steps it takes to go from the smaller number to the larger. By cunningly planning his course of action, the child realizes a remarkable economy. It takes him the same number of steps—only two—to compute 8–2 and 8–6. But how does he select the appropriate strategy? The optimal choice is dictated by the size of the number to be subtracted. If it is greater than half the starting number, as in 8–5, 8–6, or 8–7, the second strategy is the winner; otherwise, as in 8–1, 8–2, or 8–3, backward counting is faster. Not only is the child a sufficiently clever mathematician to spontaneously discover this rule, but he manages to use his natural sense of numerical quantities to apply it. The selection of an exact calculation strategy is guided by an initial quick guess. Between the age of four and seven, children exhibit an intuitive understanding of what calculations mean and how they should best be selected.

Memory Appears on the Scene

Take a stopwatch and measure how long a seven-year-old child takes to add two numbers. You will discover that the calculation time increases in direct proportion to the smaller addend, a sure sign that the child is using the minimum algorithm. Even if the child betrays no evidence of counting either verbally or on his fingers, response times indicate that he is reciting the numbers in his head. Computing 5+1, 5+2, 5+3, or 5+4 takes him an additional four-tenths of a second for each additional unit: At that age, each counting step takes about 400 milliseconds.

What happens in older subjects? When they first conducted this experiment in 1972, Carnegie-Mellon University psychologist Guy Groen and his student John Parkman were puzzled to discover that even in college students, the duration of an addition is predicted by the size of the smaller addend. The only difference is that the size of the time increment is much smaller: 20 milliseconds per unit. How should this finding be interpreted? Surely even talented students cannot count at the incredible speed of 20 milliseconds per digit, or 50 digits per second. Groen and Parkman thus proposed a hybrid model. On 95% of trials, the students would directly retrieve the result from memory. On the remaining 5% of trials, their memory would collapse, and they would have to count at the speed of 400 milliseconds per digit. On average, therefore, addition times would increase by only 20 milliseconds for each unit.

Despite its ingenuity, this proposal was quickly challenged by new findings. It was soon realized that students' response time did not increase linearly with the size of the addends (Figure 5.1). Large addition problems such as 8+9 took a disproportionately long time. The time to add two digits was actually best predicted

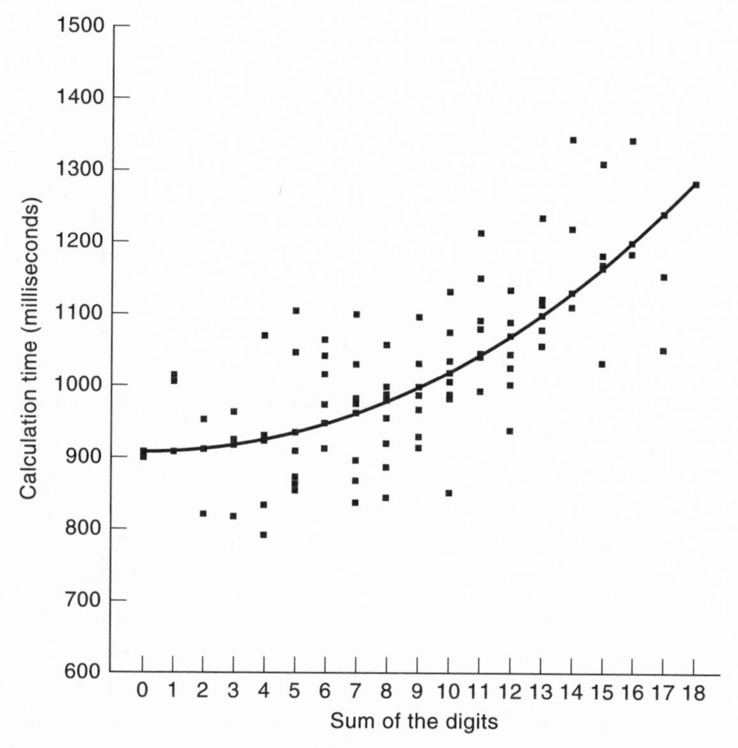

Figure 5.1. The problem-size effect: The time for an adult to solve an addition problem increases sharply with the size of the addends. (Reprinted from Ashcraft 1995 by permission of the author and publisher; copyright © 1995 by Erlbaum (UK), Taylor & Francis, Hove, UK.)

by their product or by the square of their sum—two variables that were hard to reconcile with the hypothesis that the subjects were counting. The final blow against the counting theory came when it was discovered that the time to *multiply* two digits was essentially identical to the time taken to add them. In fact, addition and multiplication times were predicted by the very same variables. If subjects counted, even on only 5% of trials, multiplication should have been much slower than addition.

There was only one way out of this conundrum. In 1978, Mark Ashcraft and his colleagues at Cleveland State University proposed that young adults hardly ever solve addition and multiplication problems by counting. Instead, they generally retrieve the result from a memorized table. Accessing this table, however, takes an increasingly longer time as the operands get larger. It takes less than a second to retrieve the result of 2+3 or 2×3, but about 1.3 seconds to solve 8+7 or 8×7.

This effect of number size on memory retrieval probably has multiple origins. As explained in previous chapters, the accuracy of our mental representation drops quickly with number size. Order of acquisition may also be a factor, because simple arithmetic facts, which involve small operands, are often learned before more difficult ones with large operands. A third factor is the amount of drilling. Because the frequency of numerals decreases with size, we receive less training with larger multiplication problems. Mark Ashcraft and his colleagues have tallied up how often each addition or multiplication problem appears in children's textbooks. The outcome is surprisingly inane: Children are drilled far more extensively with multiplications by 2 and by 3 than by 7, 8, or 9, although the latter are more difficult.

The hypothesis that memory plays a central role in adult mental arithmetic is now universally accepted. This does not imply that adults do not also have many other calculation strategies at their disposal. Indeed, most adults confess to using indirect methods such as computing 9×7 as (10×7)–7, a factor that also contributes to slowing down the resolution of large addition and multiplication problems. It does mean, however, that a major upheaval in the mental arithmetic system occurs during preschool years. Children suddenly shift from an intuitive understanding of numerical quantities, supported by simple counting strategies, to a rote learning of arithmetic. It is hardly surprising if this major turn coincides with the first serious difficulties that children encounter in mathematics. All of a sudden, progressing in mathematics means storing a wealth of numerical knowledge in memory. Most children get through as best as they can. But as we will see, they often lose their intuitions about arithmetic in the process.

The Multiplication Table: An Unnatural Practice?

Few lessons are drilled as extensively as the addition and multiplication tables. We have all spent a portion of our childhood learning them, and as adults we constantly appeal to them. Any student executes tens of elementary calculations daily. Over a lifetime, we must solve more than ten thousand multiplication problems. And yet our arithmetic memory is at best mediocre. It takes a well-trained young adult considerable time, often more than 1 second, to solve a multiplication such as 3×7. Error rates average 10 to 15 percent. On the most difficult problems such as 8×7 or 7×9, failure occurs at least once in every four attempts, often following more than 2 seconds of intense reflection.

Why is this? Multiplications by 0 or 1 obviously do not have to be learned by rote. Furthermore once 6×9 or 3+5 are stored, the responses to 9×6 and 3+5 easily follow by commutativity. This leaves us with only forty-five addition and thirty-six multiplication facts to be remembered. Why is it so difficult for us to store them? After all, hundreds of other arbitrary facts crowd our memory. The names of our friends, their ages, their addresses, and the many events of our lives occupy entire sections of our memory. At the very age when children labor over arithmetic, they effortlessly acquire a dozen new words daily. Before adulthood, they will have learned at least twenty thousand words and their pronunciation, spelling, and meaning. What makes the multiplication table so much harder to retain, even after years of training?

The answer lies in the particular structure of addition and multiplication tables. Arithmetic facts are not arbitrary and independent of each other. On the contrary, they are closely intertwined and teeming with false regularities, misleading rhymes, and confusing puns. What would happen if you had to memorize an address book that looked like this:

- Charlie David lives on George Avenue.
- Charlie George lives on Albert Zoe Avenue.
- George Ernie lives on Albert Bruno avenue.

And a second one for professional addresses like this:

- Charlie David works on Albert Bruno Avenue.
- Charlie George works on Bruno Albert Avenue.
- George Ernie works on Charlie Ernie Avenue.

Learning these twisted lists would certainly be a nightmare. Yet they are nothing but addition and multiplication tables in disguise. They were composed by replacing each of the digits 0, 1, 2, 3, 4 . . . by a surname (Zoe, Albert, Bruno, Charlie, David . . .). Home address was substituted for addition, and professional address for multiplication. The six above addresses are thus equivalent to the additions 3+4=7, 3+7=10, and 7+5=12, and to the multiplications 3×4=12, 3×7=21, and 7×5=35. Seen from this unusual angle, arithmetic tables regain for our adult eyes the intrinsic difficulties that they pose for children who first discover them. No wonder we have trouble remembering them: The most amazing thing may well be that we *do* eventually manage to memorize most of them!

We haven't quite answered our question, though: Why is this type of list so difficult to learn? Any electronic agenda with a minuscule memory of less than a kilobyte has no trouble storing them all. In fact, this computer metaphor almost begs the answer. If our brain fails to retain arithmetic facts, that is because the organization of human memory, unlike that of a computer, is *associative*: It weaves multiple links among disparate data. Associative links permit the reconstruction of memories on the basis of fragmented information. We invoke this reconstruction process, consciously or not, whenever we try to retrieve a past fact. Step by step, the perfume of Proust's madeleine evokes a universe of memories rich in sounds, visions, words, and past feelings.

Associative memory is a strength as well as a weakness. It is a strength when it enables us, starting from a vague reminiscence, to unwind a whole ball of memories that once seemed lost. No computer program to date reproduces anything close to this "addressing by content." It is a strength again when it permits us to take advantage of analogies and allows us to apply knowledge acquired under other circumstances to a novel situation. Associative memory is a weakness, however, in domains such as the multiplication table where the various pieces of knowledge must be kept from interfering with each other at all costs. When faced with a tiger, we must quickly activate our related memories of lions. But when trying to retrieve the result of 7×6, we court disaster by activating our knowledge of 7+6 or of 7×5. Unfortunately for mathematicians, our brain evolved for millions of years in an environment where the advantages of associative memory largely compensated for its drawbacks in domains like arithmetic. We are now condemned to live with inappropriate arithmetical associations that our memory recalls automatically, with little regard for our efforts to suppress them.

Proof of the pernicious influence of interference in associative memory is easy to come by. Throughout the world, scores of students have contributed hundreds of thousands of response times and tens of thousands of errors to the scientific study of calculation processes. Thanks to them, we now know precisely which calculation errors are the most frequent. Multiply 7 by 8. It is probable that instead of 56, you will answer 63, 48, or 54. Nobody ever replies 55, although this number is only one unit off the correct result. Practically all errors belong to the multiplication table, most often to the same line or column as the original multiplication problem. Why? Because the mere presentation of 7×8 is enough for us to not only recall the correct result 56, but also its tightly associated neighbors 7×9, 6×8, or 6×9. All of these facts compete in gaining access to speech production processes. All too often we try to retrieve 7×8 and the result of 6×8 pops up.

The automatization of arithmetic memory starts at a young age. As early as seven, whenever we see two digits, our brain automatically cranks up their sum. To prove this, psychologist JoAnne Lefevre and her colleagues at the University of Alberta in Canada concocted a clever experiment. They explained to subjects that they were going to see a pair of digits such as 2 and 4 that they had to memorize for a second. They would then see a third digit and were to decide whether it was identical to one of the first two numbers. The results revealed an unconscious addition process. When the target digit was equal to the sum of the pair (6), although the subjects generally responded correctly that it was not equal to any of the initial digits, there was a noticeable slowing of responses, which was not seen for neutral targets such as 5 or 7. In a recent study by Patrick Lemaire and collaborators, this effect was replicated with children as young as seven. Apparently, the mere flashing of the digits 2 and 4, even without a plus sign, suffices for our memory to automatically retrieve their sum. Subsequently, because this number is active in our memory, we are not quite sure whether we have seen it or not.

Here is another striking demonstration of the automaticity of arithmetic memory that you can try for yourself. Answer the following questions *as fast as you can*:

- 2+2?
- 4+4?
- 8+8?
- 16+16?

Now quick! Pick a number between 12 and 5. Got it?

The number you picked is 7, isn't it?

How did I read your mind? The mere presentation of the numbers 12 and 5 seems enough to trigger an unconscious subtraction 12–5=7. This effect is probably amplified by the initial addition drill, the reversed order of the numbers 12 and 5, and the ambiguous phrase "between 12 and 5" that may incite you to compute the distance between the two numbers. All these factors conspire to enhance the automatic activation of 12–5 up to a point where its result enters consciousness. And you believed that you were exercising your "free will" when selecting a digit!

Our memory also has a hard time keeping addition and multiplication facts in distinct compartments. Not infrequently do we automatically answer an addition problem with the corresponding multiplication fact (2+3=6); more rarely, the contrary occurs (3×3=6). It also takes us longer to realize that 2×3=5 is false than to reject 2×3=7 because the former result would be correct under addition.

Kevin Miller, at the University of Texas, has studied how such interference evolves during the acquisition of new arithmetic facts. In third grade, most pupils already know many additions by heart. As they start to learn multiplication, the time they take to solve an addition temporarily *increases*, while the first memory slips of the 2+3=6 kind begin to appear. Thus, the integration of multiple arithmetic facts in long-term memory seems to be a major hurdle for most children.

Verbal Memory to the Rescue

If storing arithmetic tables in memory is so difficult, how does our brain eventually catch up? A classic strategy consists in recording arithmetic facts in verbal memory. "Three times seven, twenty-one" can be stored word for word alongside "Twinkle twinkle little star" or "Our Father who art in Heaven." This solution is not unreasonable because verbal memory is vast and durable. Indeed, who does not still have a head full of slogans and songs heard years earlier? Educators have long realized the huge potential of verbal memory. In many countries, recitation remains the prime method for teaching arithmetic. I still remember the ungracious chorus at elementary school as my fellow budding mathematicians and I loudly recited multiplication tables in perfect synchrony.

The Japanese seem to have pushed this method even further. Their multiplication table is made up of little verses called "ku-ku." This word, which literally means "nine-nine," is directly drawn from the last verse of the table, 9×9=81. In the Japanese table, times and equal symbols are silent, leaving only the two operands and the result. Thus 2×3=6 is learned as "ni san na-roku"—literally, "two three zero six." Several conventions have been consecrated by history. In ku-ku, numbers are pronounced in their Chinese form, and their pronunciation varies with context. For instance, eight is normally "hashi," but can be abbreviated as "hap" or even as "pa," as in "hap-pa roku-ju shi," 8×8=64. The resulting system is complex and often arbitrary, but its singularities probably ease the load on memory.

The fact that arithmetic tables are learned verbatim seems to have an intriguing consequence: Calculation becomes tied to the language in which it is learned at school. An Italian colleague of mine, after spending more than twenty years in the United States, is now an accomplished bilingual. He speaks and writes in fluent English, with a rigorous syntax and an extensive vocabulary. Yet when he has to calculate mentally, he can still be heard mumbling numbers in his native Italian. Does this mean that after a certain age, the brain loses its plasticity for learning arithmetic? This is a possibility, but the real explanation may be more trivial. Learning arithmetic tables is so laborious that it may be more economical for a bilingual to switch back to the mother tongue for calculation rather than relearn arithmetic from scratch in a new language.

Non-bilinguals can experience the same phenomenon. We all find it hard to refrain from naming numbers aloud when we have to perform complex calculations. The crucial role played by the verbal code in arithmetic becomes fully apparent when one is asked to calculate while simultaneously reciting the alphabet aloud. Try it, and you will easily convince yourself that this is quite hard, because speaking saturates the cerebral language production systems necessary for mental calculation.

Yet a better proof of the verbatim coding of the multiplication table comes from the study of calculation errors. When confronted with 5×6, we often mistakenly respond "36" or even "56," as if the 5 and the 6 of the problem contaminated our response. Our cerebral circuits tend to automatically read the problem as a two-digit number: 5×6 irrepressibly evokes the words "fifty-six." Most strangely, this reading bias interacts in a complex way with the plausibility of the result. One never observes gross blunders such as 6×2=62 or 3×7=37. Most of the time, we mistakenly read the operands only when the resulting two-digit number is a plausible result that belongs to the multiplication table (for instance, 3×6=36 or 2×8=28). This suggests that reading errors do not occur *after* multiplication retrieval, but *during* it—at a time when the reading bias can still influence access to arithmetic memory without completely overriding it. Hence, reading and arithmetic memory are highly interconnected procedures that make use of the same verbal encoding of numbers. For the adult brain, multiplying merely means reading out 3×6 as "eighteen."

In spite of its importance, verbal memory is not the only source of knowledge to be exploited during mental calculation. When confronted with the difficult task of memorizing arithmetic tables, our brain uses every available artifice. When memory fails, it falls back on other strategies like counting, serial addition, or subtraction from some reference (for instance, 8×9=(8×10)–8=72). Above all, it never misses any opportunity to take a shortcut. Please verify whether the following calculations are true or false: 5×3=15, 6×5=25, 7×9=20. Do you have to calculate to reject the third multiplication? Probably not, for at least two good reasons. First, the proposed result, 20, is grossly false. Experiments have shown that response time drops as the degree of falsehood increases. Results whose magnitude departs considerably from the truth are rejected in less time than it would take to actually complete the operation, suggesting that in parallel to calculating the exact result, our brain also computes a coarse estimate of its size. Second, in 7×9=20, parity is violated. Since both operands are odd, the result should be odd. An analysis of response times shows that our brain implicitly checks the parity rules that govern addition and multiplication and quickly reacts whenever a violation is found.

Mental Bugs

Let us now briefly tackle the issue of multidigit calculations. Suppose that you have to compute 24+59. No computer would need more than a few microseconds, yet it will take you more than two seconds, or at least a hundred thousand times longer. This problem will mobilize all your power of concentration (as we will see later on, the prefrontal sectors of the brain, which are involved in the control of nonautomated activities, are highly active during complex calculations). You will have to go carefully through a series of steps: Isolate the rightmost digits (4 and 9), add them up (4+9=13), write down the 3, carry the 1, isolate the leftmost digits (2 and 5), add them up (2+5=7), add the carry over (7+1=8), and finally write down the 8. These stages are so reproducible that given the magnitude of the digits, one can estimate the duration of each operation and predict, to within a few tens of a second, at which point you will finally lift your pen.

At no time during such a calculation does the meaning of the unfolding operations seem to be taken into account. Why did you carry the 1 over to the leftmost column? Perhaps you now realize that this 1 stands for 10 units and that it must therefore land in the tens column. Yet this thought never crossed your mind while you were computing. In order to calculate fast, the brain is forced to ignore the meaning of the computations it performs.

As another example of the divorce between the mechanical aspects of calculation and their meaning, consider the following subtraction problems, which are quite typical of a young child:

54	54	612	317
−23	−28	− 39	− 81
31	34	627	376
(correct)	(false)	(false)	(false)

Do you see the problem? This child is not responding at random. Every single answer obeys the strictest logic. The classical subtraction algorithm is rigorously applied, digit after digit, from right to left. The child, however, reaches an impasse whenever the top digit is smaller than the bottom. This situation calls for carrying over, but for some reason the child prefers to invert the operation and subtract the top digit from the bottom one. Little does it matter that this operation is meaningless. Indeed, the result often exceeds the starting number, without disturbing the pupil in the least. Calculation appears to him as a pure manipulation of symbols, a surrealist game largely devoid of meaning.

John Brown, Richard Burton, and Kurt Van Lehn, from Carnegie-Mellon University, studied mental subtraction with such meticulous scrutiny that they wound up collecting the responses of more than a thousand children to tens of problems. In this way, they discovered and classified dozens of systematic errors similar to the ones we've just examined. Some children have difficulties only with zeroes, while others fail only with the digit 1. A classical error consists in a leftward shift of all carry-overs that apply to the digit 0. In 307–9, some children correctly compute 17–9 = 8, but then fail to subtract the carryover from 0. Instead, they wrongly simplify the task by carrying over the 1 into the hundreds column; "therefore", 307–9 = 208. Errors of this kind are so reproducible that Brown and his colleagues have described them in computer science terms: Children's subtraction algorithms are riddled with "bugs."

Where do these bugs come from? Strange as it might seem, no textbook ever describes the correct subtraction recipe in its full generality. A computer scientist can vainly search his kid's arithmetic manual for instructions precise enough to program a general subtraction routine. All school manuals are content with providing rudimentary instructions and a panoply of examples. Pupils are supposed to study the examples, analyze the behavior of their teacher, and derive their own conclusions. It is hardly surprising, then, that the algorithm they arrive at is not correct. Textbook examples generally do not cover all possible cases of subtraction. Hence they leave the door open to all sorts of ambiguities. In due course, any child is confronted with a novel situation where he or she will have to improvise, and gaps in his or her understanding of subtraction will show up.

Consider this example studied by Kurt Van Lehn: A child subtracts correctly except that each time he has to subtract two identical digits, he wrongly carries 1 over to the next column (e.g., 54–4 = 40; 428–26 = 302). This child has correctly figured out that one must carry over whenever the top digit is smaller than the bottom. However, he wrongly generalizes this rule to the case where the two digits are equal. Most likely, this particular case was never dealt with in his textbook.

Another edifying example: Many arithmetic textbooks illustrate only the subtraction procedure with two-digit numerals (17–8, 54–6, 64–38, etc.). Initially then, pupils only learn to carry over to the tens column, which is always the first column from the left. Hence the first time they are confronted with a three-digit subtraction, many children wrongly decide to carry over to the leftmost column as they have in the past (e.g., 621–2 = 529). How could they guess, without further instruction, that one should always carry over from the column *immediately left of the present one,* rather than from the leftmost column? Only a refined understanding of the algorithm's design and purpose can help. Yet the very occurrence of such absurd errors suggests that the child's brain registers and executes most calculation algorithms without caring much about their meaning.

Pros and Cons of the Electronic Calculator

What coherent picture emerges from this panorama of human arithmetic abilities? Clearly, the human brain behaves unlike any computer that we currently know of. It has not evolved for the purpose of formal calculation. This is why sophisticated arithmetic algorithms are so difficult for us to faithfully acquire and execute. Counting is easy because it exploits our fundamental biological skills for verbal recitation and one-to-one correspondence. But memorizing the multiplication table, executing the subtraction algorithm, and dealing with carryovers are purely formal operations without any counterpart in a primate's life. Evolution can hardly have prepared us for them. The *Homo sapiens* brain is to formal calculation what the wing of the prehistoric bird *Archaeopteryx* was to flying: a clumsy organ, functional but far from optimal. To comply with the requirements of mental arithmetic, our brain has to tinker with whatever circuits it has, even if that implies memorizing a sequence of operations that one does not understand.

We cannot hope to alter the architecture of our brain, but we can perhaps adapt our teaching methods to the constraints of our biology. Since arithmetic tables and calculation algorithms are, in a way, counternatural, I believe that we should seriously ponder the necessity of inculcating them in our children. Luckily, we now have an alternative—the electronic calculator, which is cheap, omnipresent, and infallible. Computers are transforming our universe to such an extent that we cannot confine ourselves thoughtlessly to the educational recipes of yesteryear. We have to face this question: Should our pupils still have to spend hundreds of hours reciting multiplication tables, as their grandparents did, in the hope that arithmetic facts will eventually be engraved in their memories? Would it not be wiser to give them early training in electronic calculators and computers?

Reducing the part played by rote arithmetic at school may be judged a heresy. Yet there is nothing sacred in the way arithmetic is currently taught. Until recently, in many countries, the abacus and finger counting were the privileged vectors of arithmetic. Even today, millions of Asians pull out their "soro-ban," the Japanese abacus, whenever they have to calculate. The most experienced of them practice the "mental abacus": By visualizing abacus moves in their heads, they can add two numbers mentally in less time than it takes us to type them into a calculator! These examples show that there are alternatives to the rote learning of arithmetic.

One might object that electronic calculators atrophy children's mathematical intuitions. This opinion has been vehemently defended, for instance, by the famous French mathematician and Fields Medal winner René Thom, who wrote,

"In primary school we learned the addition and multiplication tables. It was a good thing! I am convinced that when children as young as six or seven are allowed to use a calculator, they eventually attain a less intimate knowledge of number than the one we reached through the practice of mental calculation."

Yet what may have been true for schoolboy Thom need not hold for the average child today. Anyone can judge for himself the purported ability of our schools to teach an "intimate knowledge of number." When a pupil readily concludes, without batting an eyelid, that 317–81 is 376, perhaps there is something rotten in the educational kingdom.

I am convinced that by releasing children from the tedious and mechanical constraints of calculation, the calculator can help them to concentrate on meaning. It allows them to sharpen their natural sense of approximation by offering them thousands of arithmetic examples. By studying a calculator's results, children can discover that subtraction always yields a result smaller than the starting number, that multiplying by a three-digit number always increases the size of the starting number by two or three digits, and thousands of similar facts. The mere observation of a calculator's behavior is an excellent way of developing number sense.

The calculator is like a road map for the number line. Give a calculator to a five-year-old, and you will teach him how to make friends with numbers instead of despising them. There are so many fascinating regularities to be discovered about arithmetic. Even the most elementary of them looks like pure magic to children. Multiplying by 10 adds a zero on the right. Multiplying by 11 duplicates a digit (2×11=22, 3×11=33, etc.). Multiplying by 3, then by 37, makes three copies of it (9×3×37=999). Can you figure out why? Because these childish examples might leave mathematically advanced readers unsatisfied, here are some more sophisticated ones:

- 11×11=121; 111×111=12321; 1111×1111=1234321; and so on. Do you see why?
- 12345679×9 = 111111111. Why? Note that the 8 is lacking!
- 11–3×3 = 2; 1111–33×33 = 22; 111111–333×333 = 222; and so on. Prove it!
- 1+2 = 3; 4+5+6 = 7+8; 9+10+11+12 = 13+14+15; and so on. Can you find a simple proof?

Do you find these arithmetic games barren and dull? Do not forget that before the age of six or seven, children do not yet despise mathematics. Everything that looks mysterious and excites their imagination feels like a game to them. They are open and ready to develop a passion for numbers if only one were willing to show them how magical arithmetic can be. Electronic calculators, as well as mathematical software for children, hold the promise of initiating them to the beauty of mathematics; a role that teachers, all too occupied in teaching the mechanics of calculation, often do not accomplish.

This being said, can and should the calculator serve as substitute to rote mental arithmetic? It would be foolish to pretend that I have the definitive answer. Reaching for a pocket calculator in order to compute 2×3 is obviously absurd, but no one is pushing toward such extremes. Yet it should be acknowledged that today the vast majority of adults never perform a multidigit calculation without resorting to electronics. Whether we like it or not, division and subtraction algorithms are endangered species quickly disappearing from our everyday lives—except in schools, where we still tolerate their quiet oppression.

At the very least, using calculators in school should lose its taboo status. Mathematics curricula are not immutable, much less perfect. Their sole objective should be to improve children's fluency in arithmetic, not perpetuate a ritual. Calculators and computers are only a few of the promising paths that educators have begun to explore. Perhaps we should study the teaching methods used in China or Japan in a less condescending manner. Recent studies by psychologists Harold Stevenson, from the University of Michigan, and Jim Stigler, from UCLA, suggest that these methods are often superior in many ways to those used in most Western countries. Just consider this simple example: In the West, we generally learn multiplication tables line after line, starting with the "times two" facts and ending with the "times nine" facts, for a total of 72 facts to be remembered. In China, children are explicitly taught to reorder multiplications by placing the smallest digit first. This elementary trick, which avoids relearning 9×6 when one already knows 6×9, cuts the amount of information to be learned by almost one half. It has a notable impact on calculation speed and error rates of Chinese pupils. Obviously, we do not have the monopoly on a well-conceived curriculum. Let us keep our eyes open to all potential improvements, whether they come from computer science or psychology.

Innumeracy: Clear and Present Danger?

In the Western educational system, children spend much time learning the mechanics of arithmetic. Yet there is a growing suspicion that many of them reach adulthood without having really understood when to apply this knowledge appropriately. Lacking any deep understanding of arithmetic principles, they are at risk of becoming little calculating machines that compute but do not think. John Paulos has given their plight a name: *innumeracy*, the analogue of illiteracy in the arithmetical domain. Innumerates are prompt in drawing hazardous conclusions based on a reasoning that is mathematical only in appearance. Here are a few examples:

- $\frac{1}{5} + \frac{2}{5} = \frac{3}{10}$ because 1+2 = 3 and 5+5 = 10.
- 0.2 + 4 = 0.6 because 4+2 = 6.
- 0.25 is greater than 0.5 because 25 is greater than 5.
- A basin of water at 35° C, plus another basin of water at 35° C, makes for a tub of very hot water at 70° C (stated by my six-year-old son).
- The temperature is in the 80s today, twice as warm as last night, when the temperature was 40° F.
- There is a 50% chance of rain for Saturday, and also a 50% chance of rain for Sunday, so there is a 100% certainty that it will rain over the weekend (heard on the local news by John Paulos).
- One meter equals 100 centimeters. Since the square root of 1 is 1, and the square root of 100 is 10, shouldn't one conclude that 1 meter equals 10 centimeters?
- Mrs. X is alarmed: the new cancer test that she took was positive. Her doctor certifies that the test is highly reliable and reads positive in 98% of cancer cases. So Mrs. X is 98% certain of having cancer. Right? [Wrong. The available information supports absolutely *no* conclusion. Suppose that only one person in 10,000 ever develops this type of cancer, and that the test yields a 5% rate of false positives. Of 10,000 people taking the test, about 500 will test positive, but only one of them will really suffer from cancer. In that case, despite her results, Mrs. X still only has one chance in 500 of developing cancer.]

In the United States, innumeracy has been promoted as a cause for national concern. Alarming reports suggest that, as early as preschool, American children lag way behind their Chinese and Japanese peers. Some educators view this "learning gap" as a potential threat to American supremacy in science and technology. The designated culprit is the educational system, its mediocre organization, and the poor training of its teachers. On the French side of the Atlantic, about every other year a similar controversy announces a new drop in children's mathematical achievement.

A French mathematics educator, Stella Baruk, has shrewdly analyzed the share of responsibility that is borne by the educational system in children's mathematical difficulties. Her favorite example is the following Monty Pythonesque problem: "Twelve sheep and thirteen goats are on a boat. How old is the captain?" Believe it or not, this problem was officially presented to French first- and second-graders in an official survey, and a large proportion of them earnestly responded "Twenty-five years, because 12+13=25"—an amazing example of innumeracy!

Though there are serious reasons for being concerned by the widespread incompetence in mathematics, my own belief is that our school system is not the only one to blame. Innumeracy has much deeper roots: Ultimately, it reflects the human brain's struggle for storing arithmetical knowledge. There are obviously many degrees of innumeracy, from the young child who thinks that temperatures can be added to the medical student who fails to compute a conditional probability. Yet all such errors share one feature: Their victims directly jump to conclusions without considering the relevance of the computations they perform. This is an unfortunate counterpart to the automatization of mental calculation. We become so skillful at the mechanics of calculation that arithmetic operations sometimes start automatically in our heads. Check your reflexes on the following problems:

- A farmer has eight cows. All but five die. How many cows remain?
- Judy owns five dolls, which is two fewer than Cathy. How many dolls does Cathy have?

Did you feel an impulse to answer "three" to both problems? The mere presentation of the words "fewer than" or "all but" suffices to trigger an automatic subtraction scheme in our minds. We have to fight against this automatism. A conscious effort is needed to analyze the meaning of each problem and form a mental model of the situation. Only then do we realize that we should *repeat* the number 5 in the first problem, and *add* 5 and 2 in the second problem. The inhibition of the subtraction scheme mobilizes the anterior portion of the brain, a region called the prefrontal cortex, which is involved in implementing and controlling nonroutine strategies. Because the prefrontal cortex matures very slowly —at least up to puberty and probably beyond—children and adolescents are most vulnerable to arithmetical impulsiveness. Their prefrontal cortical areas have not yet had much opportunity to acquire the large repertoire of refined control strategies required to avoid falling into arithmetic traps.

My hypothesis, then, is that innumeracy results from the difficulty of controlling the activation of arithmetic schemas distributed in multiple cerebral areas. As we shall see in Chapters 7 and 8, number knowledge does not rest on a single specialized brain area, but on vast distributed networks of neurons, each performing its own simple, automated, and independent computation. We are born with an "accumulator circuit" that endows us with intuitions about numerical quantities.

With language acquisition, several other circuits that specialize in the manipulation of number symbols and in verbal counting come into play. The learning of multiplication tables recruits yet another circuit specialized for rote verbal memory; and the list could probably go on for a long while. Innumeracy occurs because these multiple circuits often respond autonomously and in a disconcerted fashion. Their arbitration, under the command of the prefrontal cortex, is often slow to emerge. Children are left at the mercy of their arithmetical reflexes. Regardless of whether they are learning to count or to subtract, they focus on calculation routines and fail to draw appropriate links with their quantitative number sense. And so innumeracy sets in.

Teaching Number Sense

If my hypothesis is correct, innumeracy is with us for a long time, because it reflects one of the fundamental properties of our brain: its modularity, the compartmentalization of mathematical knowledge within multiple partially autonomous circuits. In order to become proficient in mathematics, one must go beyond these compartmentalized modules and establish a series of flexible links among them. The numerical illiterate performs calculations by reflex, haphazardly and without any deep understanding. The expert calculator, on the contrary, juggles mentally with number notations, moves fluently from digits to words to quantities, and thoughtfully selects the most appropriate algorithm for the problem at hand.

From this perspective, schooling plays a crucial role not so much because it teaches children new arithmetic techniques, but also because it helps them draw links between the mechanics of calculation and its meaning. A good teacher is an alchemist who gives a fundamentally modular human brain the semblance of an interactive network. Unfortunately, our schools often do not quite meet this challenge. All too often, far from smoothing out the difficulties raised by mental calculation, our educational system increases them. The flame of mathematical intuition is only flickering in the child's mind; it needs to be fortified and sustained before it can illuminate all arithmetic activities. But our schools are often content with inculcating meaningless and mechanical arithmetical recipes into children.

This state of affairs is all the more regrettable because, as we have seen, most children enter preschool with a well-developed understanding of approximation and counting. In most math courses, this informal baggage is treated as a handicap rather than as an asset. Finger counting is considered a childish activity that a good education will quickly do away with. How many children try to hide when they count on their fingers because "the teacher said not to"? Yet the history of numeration systems repeatedly proves that finger counting is an important precursor to learning base 10. Likewise, failing to retrieve 6+7=13 from rote memory is considered an error, even if the child later proves his or her excellent command of arithmetic by recovering the result indirectly—for instance, by remembering that 6+6 is 12 and that 7 is one unit after 6. Blaming a child for calling on indirect strategies blatantly ignores that adults use similar strategies when their memory fails.

Despising children's precocious abilities can have a disastrous effect on their subsequent opinion of mathematics. It accredits the idea that mathematics is an arid domain, detached from intuition and ruled by arbitrariness. Pupils feel that they are supposed to do as the teacher does, whether or not they can make any sense of it. A random example: Developmental psychologist Jeffrey Bisanz asked six- and nine-year-old pupils to calculate 5+3–3. The six-year-olds often responded 5 without calculating, rightly noting that +3 and –3 cancel each other. However the nine-years-olds, although they were more experienced, stubbornly performed the calculation in full (5+3=8, then 8–3=5). "It would be cheating to take shortcuts," explained one of them.

The insistence on mechanical computation at the expense of meaning is reminiscent of the heated debate that divides the formalist and intuitionist schools of mathematical research. The formalist trend, which was founded by Hilbert and was pursued by major French mathematicians grouped under the pseudonym of Bourbaki, set as its goal the anchoring of mathematics on a firm axiomatic base. Their objective was to reduce demonstration to a purely formal manipulation of abstract symbols. From this arid vision stemmed the all-too-famous reform of "modern mathematics," which ruined the mathematical sense of a generation of French pupils by presenting, according to an actor of this period, "an extremely formal education, cut from any intuitive support, presented on the basis of artificial situations, and highly selective." For instance, the reformers thought that children should be familiar with the general theoretical principles of numeration before being taught the specifics of our base-10 system. Hence, believe it or not, some arithmetic textbooks started off by explaining that 3+4 is 12 —in base 5! It is hard to think of a better way to befuddle children's thinking.

This erroneous conception of the brain and of mathematics, in which intuition is discouraged, leads to failure. Studies conducted in the United States by David Geary and his colleagues at the University of Missouri–Columbia indicate that about 6% of pupils are "mathematically disabled." I cannot believe that a genuine neurological handicap affects that many children. Although cerebral lesions can selectively impair mental calculation, as we will see in Chapter 7, they are relatively infrequent. It seems more likely that many of these "mathematically disabled" children are normally abled pupils who got off to a false start in mathematics. Their initial experience unfortunately convinces them that arithmetic is a purely scholastic affair, with no practical goal and no obvious meaning. They rapidly decide that they will never be able to understand a word about it. The already considerable difficulties posed by arithmetic to any normally constituted brain are thus compounded by an emotional component, a growing anxiety or phobia about mathematics.

We can fight these difficulties if we ground mathematical knowledge on concrete situations rather than on abstract concepts. We need to help children realize that mathematical operations have an intuitive meaning, which they can represent using their innate sense of numerical quantities. In brief, we must help them build a rich repertoire of "mental models" of arithmetic. Consider the example of an elementary subtraction, 9–3=6. As adults, we know of many concrete situations to which this operation applies: a set scheme (a basket containing nine apples, from which one takes away three apples, now only has six), a distance scheme (in any board game, in order to move from cell 3 to cell 9, six moves are required), a temperature scheme (if it is 9 degrees and the temperature drops 3 degrees, then it will be only 6 degrees), and many others. All such mental models seem equivalent to our adult eyes, but they are not so to the child who must discover that subtraction is the operation suited to all of them. The day the teacher introduces negative numbers and asks pupils to compute 3–9, a child who only masters the set scheme judges this operation impossible. Taking 9 apples from 3 apples? That's absurd! Another child who relies exclusively on the distance scheme concludes that 3–9=6, because indeed the distance from 3 to 9 is 6. If the teacher merely maintains that 3–9 equals "minus six," the two children run the risk of failing to understand the statement. The temperature scheme, however, can provide them with an intuitive picture of negative numbers. Minus six degrees is a concept that even first-graders can grasp.

Consider a second example: the addition of two fractions ½ and ⅓. A child who has in mind an intuitive picture of fractions as portions of a pie—half a pie, and then another third of a pie—will have little difficulty figuring out that their sum falls just below 1. He or she may even understand that the portions must be cut into smaller identical pieces (i.e., reduced to the same denominator) before they can be regrouped in order to compute the exact total $^1/_2 + {}^1/_3 = {}^5/_6$. In contrast, a child for whom fractions have no intuitive meaning and are merely two digits separated by a horizontal bar is likely to fall into the classic trap of adding the numerator and denominator: $^1/_2 + {}^1/_3 = (1+1)/(2+3) = {}^2/_5$! This error may even be justified by a concrete model. Suppose that in the first period Michael Jordan scores once in two shots, for an average of ½, and that in second period he scores once in three shots, for an average of ⅓. Over the entire game he would have scored twice in five shots. Here is a situation in which ½ "plus" ⅓ equals ⅖! When one teaches fractions, it is vital to let the child know that one has a "portion of pie" scheme in mind rather than a "scoring average" scheme. The brain is not content with abstract symbols: Concrete intuitions and mental models play a crucial role in mathematics. This is probably why the abacus works so well for Asian children; it provides them with a very concrete and intuitive representation of numbers.

But let us leave this chapter with a note of optimism. The craze for "modern mathematics," based on a formalist vision of mathematics, is losing momentum in many countries. In the United States, the national council of teachers of mathematics is now de-emphasizing the rote learning of facts and procedures and is focusing instead on teaching an intuitive familiarity with numbers. In France—the country that was most directly struck by "Bourbakism"—many teachers no longer wait for a psychologists' advice to tell them to head back to a more concrete approach to mathematics. Schools have slowly readopted concrete educational material such as Maria Montessori's bicolored bars, Seguin's tables, unit cubes, ten bars, hundreds plaques, dice, and board games. The French Ministry of Education, after several reforms, seems to have dropped the idea of turning each schoolchild into a symbol-crunching machine. Number sense—indeed, common sense—is making a comeback.

In parallel to this welcome change, education psychologists in the United States have demonstrated empirically the merits of an arithmetic curriculum that stresses concrete, practical, and intuitive mental models of arithmetic. Sharon Griffin, Robbie Case, and Robert Siegler, three North-American developmental psychologists, have joined efforts to study the impact of different educational strategies on children's understanding of arithmetic. Their theoretical analysis, like mine, emphasizes the central role played by an intuitive representation of quantities on the mental number line. On this basis, Griffin and Case designed the "RightStart" program, an arithmetic curriculum for kindergartners that comprises entertaining numerical games with varied concrete pedagogical materials (thermometers, board games, number lines, rows of objects, etc.). Their goal was to teach children from low-income inner-city neighborhoods the rudiments of arithmetic: "The central objective of the program is to enable children to relate the world of numbers to the world of quantity and consequently, to understand that numbers have meaning and can be used to predict, to explain, and to make sense of the real world."

Most children spontaneously understand the correspondence between numbers and quantities. Underprivileged children, however, may not have grasped it before entering preschool. Lacking the conceptual prerequisites for learning arithmetic, they run the risk of losing ground in mathematics courses. The

RightStart program attempts to set them back on the right path using simple interactive arithmetic games. For example, in one section of the program, children are invited to play a simple board game that teaches them to count their moves, to subtract in order to find out how far they are from the goal, and to compare numbers in order to discover who is closest to winning the game.

The results are remarkable. Griffin, Case, and Siegler have tried their program in several inner-city schools in Canada and the United States, mostly with immigrant children from low-income families. Children who were lagging behind their peers participated in forty 20-minute sessions of the RightStart program and were propelled to the top of their class as of the next semester. They even outranked pupils with a better initial command of arithmetic but who had followed a more traditional curriculum. Their advance was consolidated in the next school year. This extraordinary success story should bring some consolation to the teachers and parents who feel that their children are allergic to mathematics. In fact, most children are only too pleased to learn mathematics if only one shows them the playful aspects before the abstract symbolism. Playing snakes and ladders may be all children need to get a head start in arithmetic.

Chapter 6

Geniuses and Prodigies

An expert is a man who has stopped thinking—he knows!
Frank Lloyd Wright

One of the most romantic episodes in the history of mathematics occurred one morning in January 1913, when professor G. H. Hardy received a strange-looking letter from India. At thirty-six, Hardy was a renowned mathematician, probably England's most brilliant. Professor at Trinity College in Cambridge, he had recently been elected a fellow of the Royal Society. There he often conversed on equal terms with minds as remarkable as Whitehead and Russell. So one can imagine his growing irritation as he skimmed through this letter posted in Madras. In rudimentary syntax, an unknown Indian named Srinavasa Ramanujan Iyengar requested his opinion on several theorems.

Despite his unforgiving contempt for amateur mathematicians, Hardy quickly became fascinated as he began to decipher with increasing attention his correspondent's mysterious mathematical formulas (Figure 6.1). Some were long-established theorems—but why on earth did the man present them as if they were his? Others were derived, sometimes via indirect routes, from deep mathematical results that Hardy knew very well for having personally contributed to them. The last few formulas, however, were unheard of, long strings of square roots, exponentials, and continuous fractions mixed in a unique cocktail whose origins remained incomprehensible.

Never had Hardy seen anything like this. It could not be a hoax: He was assuredly confronted with a first-rate genius. As he later explained in his autobiography, "The formulas had to be true because, if they were not, no one would have had the imagination to invent them." The following day, Hardy resolved to help Ramanujan come to Cambridge. This was the starting point of an extremely fertile collaboration that culminated with Ramanujan's election to the Royal Society a few years later, and ended tragically with his death on April 26, 1920, at the age of 32.

One could argue, with only a pinch of irony, that Ramanujan's genius overran Isaac Newton's, because he had seen further than any other mathematician without sitting on anybody's shoulders. Born to a poor Brahmin family, Ramanujan had received only nine years of study at Kumbakonam's local school, in South India, and had never obtained a university degree. Early on in his childhood, however, his genius was already apparent. He had rediscovered on his own the famous Euler formulas that link trigonometric and exponential functions, and by the time he was twelve he had already mastered S. Loney's *Plane Trigonometry*.

At sixteen, Ramanujan encountered a second book that decided his mathematical bent. It was G. S. Carr's *Synopsis of Elementary Results in Pure and Applied Mathematics*—a compilation of 6,165 theorems with only sketchy demonstrations. By dint of studying this austere volume and reinventing for himself the

$$\frac{2}{\pi} = 1 - \left(\frac{1}{2}\right)^3 + 9\left(\frac{1}{2}\times\frac{3}{4}\right)^3 - 13\left(\frac{1}{2}\times\frac{3}{4}\times\frac{5}{6}\right)^3 + 17\left(\frac{1}{2}\times\frac{3}{4}\times\frac{5}{6}\times\frac{7}{8}\right)^3 \cdots$$

$$\cfrac{1}{1+\cfrac{e^{-2\pi\sqrt{5}}}{1+\cfrac{e^{-4\pi\sqrt{5}}}{1+\cfrac{e^{-6\pi\sqrt{5}}}{1+\ddots}}}} = \left[\frac{\sqrt{5}}{1+\sqrt[5]{5^{3/4}\left(\frac{\sqrt{5}-1}{2}\right)^{5/2}-1}} - \frac{\sqrt{5}+1}{2}\right] e^{2\pi\sqrt{5}}$$

$$\pi \cong \frac{-2}{\sqrt{210}} \log\left[\frac{(\sqrt{2}-1)^2(2-\sqrt{3})(\sqrt{7}-\sqrt{6})^2(8-3\sqrt{7})(\sqrt{10}-3)^2(\sqrt{15}-\sqrt{14})(4-\sqrt{15})^2(6-\sqrt{35})}{4}\right]$$

Figure 6.1. A small sample of Ramanujan's mysterious formulas. The bottom expression for number π is correct to twenty decimal places.

mathematics of past centuries, Ramanujan acquired a singular genius that no mathematician before or after him seems to have possessed to the same degree: an uncanny sense of the right formula, a refined intuition of numerical relations. He was unmatched in his ability to envision novel arithmetical relations that nobody had dreamt of previously, and which he generally accepted on the basis of intuition alone—to the great despair of his fellow mathematicians who, until very recently, have been at pains to provide rigorous proofs or refutations for the hundreds of formulas that filled his notebooks.

Ramanujan claimed that his theorems were "written on his tongue" during the night by the goddess Namagiri. On getting out of bed, he would often feverishly write down some unexpected result that would later stun his colleagues. Personally, I am rather skeptical about the central role played by Indian divinities at the forefront of mathematical research. But the ball is in the neuropsychologist's court—can psychology or neurology propose at least an embryo of an explanation for the extraordinary fertility of this unique mind?

Almost fifty years after Ramanujan's death, England saw the birth of another genius whose talent was, in several respects, the exact parallel and yet the opposite of Ramanujan's. Michael is a profoundly retarded autistic young man who was studied for years by two English psychologists, Beate Hermelin and Neil O'Connor. As a child, he suffered from macrocephalia and had convulsions that probably betrayed early brain damage. He was an upset and destructive child, oblivious to danger, who seemed to live in a shut-down and self-centered world. Never did he wave good-bye or point to objects—gestures that very young children normally acquire spontaneously. Never did he show any interest in the company of adults.

Although Michael is now in his twenties, he still cannot speak. He never learned sign language and shows no evidence that he understand words. His verbal IQ is not measurable in any test that requires the use of words. Even in a nonverbal test, his IQ only reaches 67. He fails essentially on all tests tapping everyday knowledge of objects.

Why compare this severely handicapped autistic man with an Indian mathematical genius? Because despite his dramatic mental retardation, Michael is extraordinarily conversant in arithmetic. Around the age of six, he learned to copy some letters and the ten Arabic digits. Since then, adding, subtracting, multiplying, dividing, and factoring numbers have been his favorite pastimes. Money, clocks, calendars, and maps also fascinate him. When measured with logical tests, his IQ reaches 128, way above the normal mean. Here is a young man who cannot name a car or a rabbit, but who immediately perceives that 627 can be decomposed into $3 \times 11 \times 19$! It takes Michael only a little over one second to determine that a three-digit number is prime (which means that it cannot be expressed as the product of two smaller numbers). A psychologist with a mathematics diploma who attempted this task took ten times longer.

How can one be mute, mentally retarded, and a lightning calculator? How can one grow up in a poor Indian family and become a top-level mathematician with only the help of two books largely devoid of demonstrations? Psychologists now know hundreds of "idiot savants" similar to Michael throughout the world. Some can tell you the day of the week for any past or future calendar date. Others can mentally add two six-digit numbers in less time than it would take us to dial them on a phone. Yet these people are often totally devoid of social intelligence and may even lack language. Does not the very existence of such prodigies jeopardize the theory that I sketched in previous chapters? How do they escape the calculation difficulties that we are all confronted with? What is the nature of the "sixth sense" that confers on them such a soundness of intuition in the numerical domain? Should we grant them a special form of cerebral organization, an innate gift for arithmetic?

A Numerical Bestiary

The role of memory in mathematics is easily underestimated. Each of us unconsciously garners hundreds of numerical facts—consider, for instance, the evocative power of the numbers 1492, 800, 911 or 2000. The size of this numerical memory store is undoubtedly one of calculating prodigies' main strengths. Their familiarity with numbers is so refined that, for them, hardly any number is random. What appears to us as an ordinary series of digits assumes a singular meaning for them. As explained by the lightning calculator G. P. Bidder: "The number 763 is represented symbolically by three figures 7-6-3; but 763 is only one quantity, one number, one idea, and it presents itself to my mind just as the word "hippopotamus" presents the idea of one animal."

Each calculating genius maintains a mental zoo peopled with a bestiary of familiar numbers. Being on familiar terms with numbers, knowing them inside out, are the hallmark of these expert arithmeticians. "Numbers are friends to me, more or less," says Wim Klein. "It doesn't mean the same for you, does it, 3,844? For you it's just a three and an eight and a four and a four. But I say: 'Hi, 62 squared!' "

Abundant biographical anecdotes confirm the extreme familiarity with which great mathematicians manipulate the tools of their trade, be they numbers or geometrical figures. The following dialog took place between Hardy and Ramanujan while the Indian mathematician was slowly dying of tuberculosis in a sanitarium. "The taxi that I hired to come here bore the number 1729," said Hardy. "It seemed a rather dull number," "Oh no, Hardy," retorted Ramanujan. "It is a captivating one. It is the smallest number that can be expressed in two different ways as a sum of two cubes"—$1{,}729 = 1^3 + 12^3 = 10^3 + 9^3$!

Gauss, another exceptional mathematician as well as a calculating prodigy, is credited with a similar performance at a young age. His teacher asked his class to add all numbers from 1 to 100, probably hoping to keep his pupils quiet for a half-hour. But little Gauss immediately raised his slate with the result. He had rapidly perceived the symmetry of the problem. By "mentally folding" the number line, he could group 100 with 1, 99 with 2, 98 with 3, and so on. Hence the sum was reduced to 50 pairs, each totaling 101, for a grand total of 5,050.

The French mathematician François Le Lionnais stresses how "the aptitudes for mental calculation and for mathematics . . . have in common a certain sensibility to what I shall call the personality of each number." In 1983, Le Lionnais published a little book called *Remarkable Numbers,* in which he listed several hundred numbers with special mathematical properties. His fascination for numbers started at the age of five. After studying the multiplication table printed on the back of his notebook, he was awed to discover that the multiples of 9 ended with the digits 9, 8, 7, 6, and so on (as in 9, 18, 27, 36, etc.; Can you see why?). As a schoolboy, a student, and eventually as a professional mathematician, he spent the rest of his life hunting for genuinely "odd" numbers and other deep mathematical results. His files were lost when he was deported to a German camp during World War II, but he started all over again from memory and added ever more gems to his collection, year after year.

In the end, his list of remarkable numbers reveals a significant amount of what a top-level mathematician must know in arithmetic. Most of his bestiary will remain forever opaque to the profane. For instance, 244,823,040, one of the few numbers to which he grants three stars, is described by him in standard mathematical language as "the order of group M_{24}, the ninth sporadic group, an example of which is the group of Steiner automorphisms with indices (5,8,24)"—a definition that leaves most of us cold! Here are some of the most accessible monuments in this Fodor's guide to the number line:

- $$\varphi = 1.618033988\ldots = \frac{1+\sqrt{5}}{2} = \sqrt{1+\sqrt{1+\sqrt{1+\sqrt{1+\ldots}}}} = 1 + \cfrac{1}{1+\cfrac{1}{1+\cfrac{1}{1+\ldots}}}$$

the famous "golden section" that supposedly underlies many works of arts, such as the Parthenon. Enter it in your pocket calculator and then press the "1/x" or "x^2" keys. The result will surprise you.

- 4: the minimum number of colors needed to color any planar map so that no two neighboring countries have the same color. Not unlike Kasparov's recent loss in chess to an IBM computer, the "four-color theorem" is famous in mathematics for marking the limits of human reasoning: Its proof calls for the successive examination of so many special cases that only a computer can complete it.
- 81: the smallest square that can be decomposed into a sum of three squares (92 = 12 + 42 + 82).
- $e^{\pi\sqrt{163}}$ a real number that falls remarkably close to an integer: Its first twelve decimals are all 9s (another of Ramanujan's contributions).
- The number formed by writing down 317 times the digit 1, which is a prime
- 1,234,567,891, also a prime
- And even 39, the smallest integer with no remarkable mathematical properties—which, as Le Lionnais himself notes, raises a paradox: Doesn't it make number 39 remarkable after all?

The Landscape of Numbers

As one browses through Le Lionnais's surrealist inventory, one cannot but think that some mathematicians must be more familiar with the number line than with their own backyard. Indeed, the metaphor of a "panorama of mathematics" seems particularly apt at capturing their vivid introspection. Most of them feel that mathematical objects have an existence of their own, as real and tangible as that of any other object. Says Ferrol, a well-known calculating prodigy: "I often feel, especially when I am alone, that I dwell in another world. Ideas of numbers take on a life of their own. Suddenly, questions of any kind rise before my eyes with their answers."

The same conception is found in the writings of the French mathematician Alain Connes: "Exploring the geography of mathematics, little by little the mathematician perceives the contours and structure of an incredibly rich world. Gradually he develops a sensitivity to the notion of simplicity that opens up access to new, wholly unsuspected regions of the mathematical landscape."

Connes thinks that expert mathematicians are endowed with a clairvoyance, a flair, a special instinct comparable to the musician's fine-tuned ear or to the wine taster's experienced palate that enables them to directly perceive mathematical objects: "The evolution of our perception of mathematical reality causes a new sense to develop, which gives us access to a reality that is neither visual, nor auditory, but something else altogether."

In *The Man Who Mistook His Wife for a Hat*, Oliver Sacks describes two autistic twins whom he once caught exchanging very large prime numbers. His interpretation also appeals to a certain "sensibility" about the mathematical world:

> They are not calculators, and their numeracy is "iconic." They summon up, they dwell among, strange scenes of numbers; they wander freely in great landscapes of numbers; they create, dramaturgically, a whole world made of numbers. They have, I believe, a most singular imagination—and not the least of its singularity is that it can imagine only numbers. They do not seem to "operate" with numbers non-iconically, like a calculator, they "see" them directly, as a vast natural scene.

For René Thom, the renowned creator of the mathematical theory of catastrophes, an intuitive perception of mathematical spaces is so essential that any mathematician who reaches the limits of his intuition feels unspeakable anxiety: "I do not feel easy with infinite dimensional spaces. I know that these are well-charted mathematical objects, whose many states are perfectly known, yet I do not like to be in a space with infinitely many dimensions. (Is it distressing?) Certainly. . . . It is a space, precisely, that defies intuition."

One can almost hear Pascal—another precocious mathematical prodigy—who stated in his *Pensées*: "The eternal silence of these infinite spaces frightens me."

The tight link between mathematical and spatial aptitudes has often been empirically demonstrated. Strong correlation exists between a person's mathematical talent and his or her scores on spatial perception tests, almost as if they were one and the same ability. Beate Hermelin and Neil O'Connor recruited a group of children between twelve and fourteen, who were judged by their teachers as particularly gifted in mathematics. They presented the children with problems that challenged their sense of spatial relations. Here is a small selection:

- How many diagonals can one draw on the surface of a cube?
- A painted wooden cube with a 9-centimeter edge is cut up into little cubes with a 3-centimeter edge each. There are thus twenty-seven such little cubes. How many of them will have only two painted sides?

Mathematically talented children were brilliant in this test. Their classmates with a standard level of achievement in mathematics, although they had an equivalent overall IQ, obtained flatly lower scores—even those that were remarkably gifted in the arts. But perhaps it is not surprising that spatial competence correlates so strongly with success in mathematics. Ever since Euclid and Pythagoras, geometry and arithmetic have been tightly linked. Establishing a spatial number map is a fundamental operation in the human brain. As we will see later on, the cerebral areas that contribute to number sense and to spatial representations occupy neighboring convolutions.

Many mathematical geniuses have claimed to possess a direct perception of mathematical relations. They say that in their most creative moments, which some describe as "illuminations," they do not reason voluntarily, nor think in words, nor perform long formal calculations. Mathematical truth descends on them, sometimes even during sleep, as in Ramanujan's case. Poincaré often declared that his intuitions convinced him of the veracity of a mathematical result, although it later took him hours of calculation to prove it formally. But it is probably Einstein himself who, in a letter published by Hadamard in his famous *Essay on the Psychology of Invention in the Mathematical Field,* articulated most clearly the role of language and intuition in mathematics: "Words and language, whether written or spoken, do not seem to play any part in my thought processes. The psychological entities that serve as building blocks for my thought are certain signs or images, more or less clear, that I can reproduce and recombine at will."

This conclusion would certainly not be challenged by Michael, the lightning-fast autistic calculating genius who lacks language. Great mathematicians' intuitions about numbers and other mathematical objects do not seem to rely so much on clever symbol manipulations as on a direct perception of significant relations. In that respect, calculating prodigies and talented mathematicians perhaps differ from the average human being only in the size of the repertoire of number facts that they can mobilize in a fraction of a second. In Chapter 3, we saw how all humans were endowed with an intuitive representation of numerical quantities, which is automatically activated whenever we see a number and which specifies that 82 is smaller than 100 without requiring any conscious effort. This "number sense" is embodied in a mental number line oriented from left to right. Only 5 percent to 10 percent of people experience it consciously as a spatial extension with varied colours and a twisted shape. Perhaps the great human calculators are one step further on this continuum. They seem to also often perceive numbers as a spatially extended domain, but with an even greater resolution and an amazing wealth of detail. In the mind of the calculating prodigy, each number does not just light up as a point on a line, but rather as an arithmetical web with links in every direction. Faced with the number 82, Ramanujan's brain instantly evokes 2×41, $100-18$, 9^2+1^2, and sundry other relations as obvious to his eyes as "smaller than 100" is to ours.

We still have to explain, however, where this prodigious intuitive memory of numbers comes from. Is it an innate gift, product of an unusual form of cerebral organization? Or does it merely result from years of training in arithmetic?

Phrenology and the Search for Biological Bases of Genius

Scientists have long been intrigued by calculating prodigies. Several theories accounting for their genius, many of them eccentric, have been put forward in the popular press. Popular candidates are gifts of God, inborn knowledge, thought transmission, or even reincarnation. Even Alfred Binet, the famous psychologist who invented the first intelligence tests, conformed to this all-out search for an explanation. In 1894, in his influential book on *The Psychology of Great Calculators and Chess Players*, which is still frequently cited, he discusses the origins of the talent of perhaps the most famous calculator of that time, Jacques Inaudi. Binet then cites, "with all the reservations that one might expect," the following anecdote:

> It appears that Inaudi's mother, while pregnant, went through psychological hardship. She watched her husband squander their meager fortune and foresaw that money would soon be lacking to face the many bills that were soon falling due. Fearing that their possessions might be seized, she computed mentally how much she should save to honour their commitments. Her days were spent buried in numbers, and she had become a calculating maniac.

Binet, being a conscientious scientist, dutifully asked himself : "Is this report accurate? And if so, could the mother's mental state have had any real influence on her son?". That Binet took this issue so seriously shows clearly how the Lamarckian theory of the inheritance of acquired traits was still very much alive in 1894, despite the publication of Darwin's *Origins of Species* in 1859.

In fact, earlier in the nineteenth century, a scientific theory of intellectual talent had already been proposed and was a recurring topic of intense discussions—the phrenological theory of mental organs. As early as 1825, Franz-Joseph Gall published his theory of "organology," later christened "phrenology" by Johann Caspar Spurzheim. His proposal clearly affirmed a materialistic vision of mind and brain that, although often ridiculed, had a profound influence on many eminent neurophysiologists, among them Paul Broca and John Hughlings Jackson. Gall's organology postulated a division of the brain into a large number of spe-

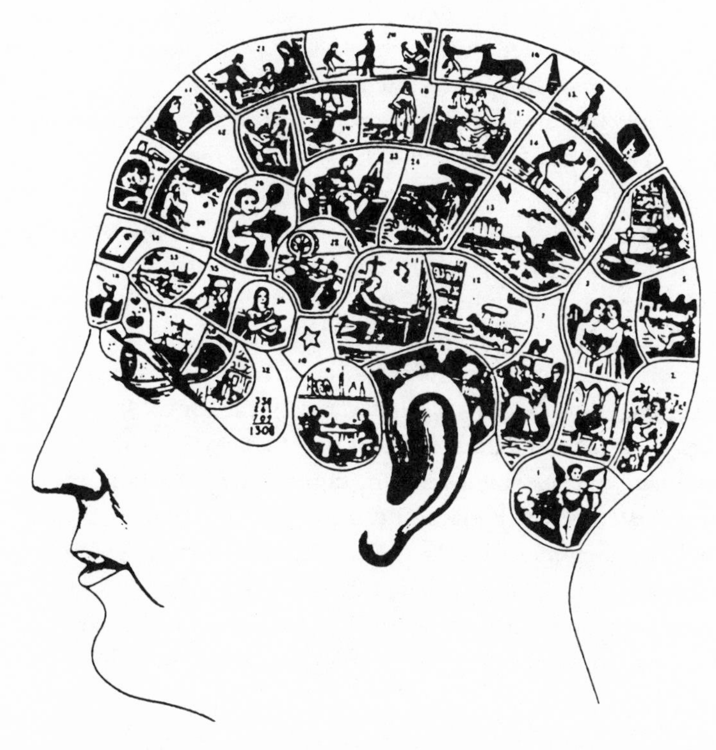

Figure 6.2. A highly figurative vision of the various cerebral organs postulated by phrenologists. The "sense of numerical relations," better known as the "math bump," was arbitrarily placed behind the eye.

cialized regions constituting as many independent innate "mental organs." Each organ supposedly subtended a precise mental faculty: the instinct of reproduction, the love of one's progeny, the memory for things and facts, the language instinct, the memory of persons, and so on. Twenty-seven faculties, which were quickly extended to thirty-five in later versions of the theory, were assigned to specific cerebral territories, often on a purely fanciful basis. In this list, the "sense of number relations" figured amid the many organs that were attributed to frontal brain areas (Figure 6.2).

Given that mental faculties were innate, how could one explain their variability from one individual to the other? Gall postulated that the relative size of cerebral organs determined each person's mental dispositions. In great mathematicians, Gall reasoned, the amount of tissue dedicated to the organ of number relations was way above average. Of course the size of cerebral convolutions was not directly accessible to measurement. But Gall proposed a simplifying assumption: The cranial bone, shaped by cortex during its growth, directly reflected the size of the underlying organs with humps and hollows. Mathematical talent could therefore be detected during childhood by "craniometry," the measurement of the deformations of the cranium. In contemporary French, a popular saying for a person highly talented in mathematics is that he or she has a "bump for math"—an expression directly inherited from phrenology.

Under the influence of Gall's theory, nineteenth-century scholars expended considerable effort on comparing the size and shape of skulls from people of different races, occupations, and intellectual levels—a scientific epic that Stephen Jay Gould has brilliantly narrated in *The Mismeasurement of Man*. Many renowned scientists fell under the spell of this fad and bequeathed their heads to science so that, in a morbid post-mortem competition, the volume of their brain matter could be compared to that of colleagues and average men. In Paris, the *société anthropologique* dedicated numerous sessions to Georges Cuvier, the famous French zoologist and paleontologist. The dimensions of his skull and even of his hat fueled a heated debate between Broca, an ardent supporter of craniometry, and Gratiolet, who contested it. Gauss's brain, which was of an average weight but was thought to have more convolutions than an ordinary German worker's brain, seemed to support Broca (Figure 6.3). Broca also noted, according to Binet, that "the young Inaudi's head was very bulky and irregular," while Charcot him-

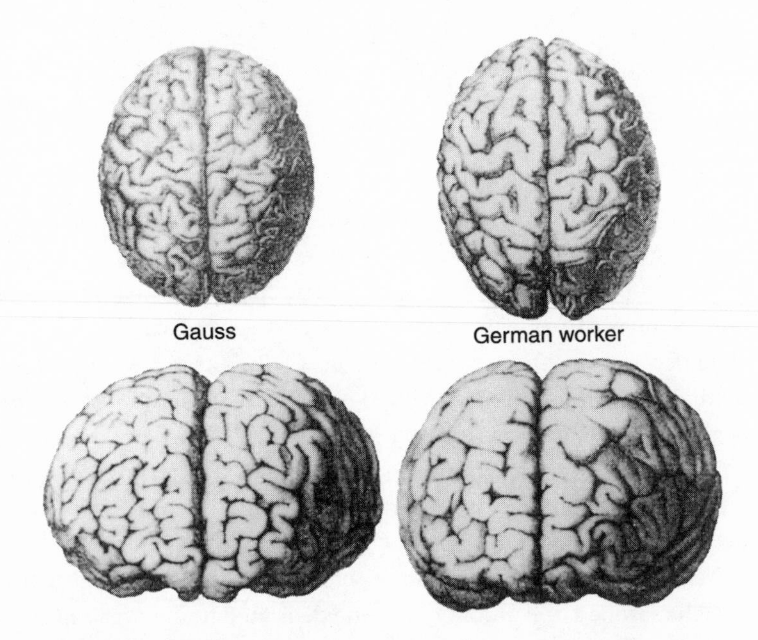

Figure 6.3. A drawing dating from the end of the nineteenth century shows many more convolutions in the brain of the genial mathematician Carl Friedrich Gauss than in an "average" German worker—an unlikely difference that probably owes more to the engraver's imagination and selection biases than to real cerebral anatomy.

self found "a slight protrusion of the right frontal hump and, on the back, a left parietal protrusion" as well as "a longitudinal crest of 0 m. 02 formed by the raised right parietal bone." The purported smaller size of the encephalon in "Negroes," women, and gorillas was interpreted as an additional proof of the tight correlation between brain size and intelligence. Needless to say, all these analyses were fraught with blatant errors that Gould, among others, has repeatedly denounced.

A century and a half later, what remains of phrenology and craniometry? Although some racists from all political sides periodically attempt to revive it, the hypothesis of a direct link between brain size and intelligence has been refuted time and again. (Gall's brain itself weighted only 1,282 grams, or 520 grams less that Cuvier's!) The legacy of Gall's organology, however, is less clear-cut. In fact, the functional specialization of cerebral areas is no longer a disputed hypothesis. It is now an established fact that every square millimeter of cortex contains neurons highly specialized for processing specific information. Indeed, we will see later how cerebral lesion studies and new methods of functional brain imaging now enable neuroscientists to draw up a sketchy map of the cerebral networks involved in mental calculation.

While these recent results undoubtedly surpass Gall's and Spurzheim's wildest dreams, they do not confirm their theory of the localization of "mental faculties." Contrary to phrenological theory, modern images of the brain never pinpoint a complex faculty such as language or calculation to a single, monolithic brain area. On contemporary maps of the brain, only very elementary functions —the recognition of a fragment of a face, the invariance of color, or the command of a motor gesture—can be assigned to a narrow cerebral region. The simplest mental act, such as reading a word, requires the orchestration of multiple assemblies of neurons distributed in diverse brain regions. It will never be possible to isolate *the* language area, even less the convolution that controls abstract thought or the region specialized in religious devotion—with all due respect to the researchers who still pursue the search for an area in charge of consciousness or altruism!

Another durable, though dubious, legacy from Gall's theory is the hypothesis that intellectual talent derives from an inborn gift, a biological predisposition to genius. In 1894, Binet thought that an "innate aptitude" accounted for the achievements of calculating prodigies. "The emergence of their faculty recalls a sort of spontaneous generation," he affirmed. Yet the study of gifted and retarded children later changed his mind. A decade later, he denied that intelligence was innate, and became an ardent supporter of special education as a means of compensating for mental retardation. To many other scientists, however, the concept of an "innate gift" was hard to kill. Even today, one of the foremost experts on "idiot savants," Neil O'Connor, perpetuates this tradition, going as far as to state that "the abilities involved [in autistic prodigies] are like innate programs of skill which come about independently of any effort of learning."

The belief that intellectual abilities are biologically determined is deeply anchored in Western thought, especially in the United States. To take just one example, psychologists Harold Stevenson and Jim Stigler have studied how American and Japanese parents rate the influence of their children's efforts versus inborn abilities in their performance at school. In Japan, the amount of effort and the quality of teaching are heralded as the most critical parameters. In the United States, on the other hand, most parents and even children themselves consider that success or failure in mathematics depends mostly on one's innate talents and limitations. A nativist bias is perceptible even in our vocabulary when we speak of talent as a "gift" (from whom?) or a "disposition" (set by whom?). The word "talented," indeed, is often considered as the opposite of "hardworking."

Until recently, even the supporters of nativist theories of intelligence scoffed at Gall's simplistic hypothesis that talent was directly proportional to the size of certain cerebral convolutions. In the last few years, however, this organological conception has made a surprising comeback in the forefront of neuroscience research. Two articles in the best international scientific journals have reported that high levels of musical competence are accompanied by an unusual extension of certain cortical areas. In musicians with perfect pitch—the ability to accurately identify the absolute pitch of a single note—a region of left-hemispheric auditory cortex named the *planum temporale* appears to be larger than that of control subjects who are devoid of this talent, regardless of whether or not they play an instrument. And in string players, the region of the sensory cortex dedicated to the tactile representation of the fingers of the left hand shows an exceptional expansion. Has musical talent been mapped?

In fact, such correlational data do not necessarily support nativist theories *à la Gall*. Studies of brain plasticity have revealed that experience may deeply modify the internal organization of brain areas. The architecture of the adult brain results from a slow process of epigenesis that extends beyond puberty and during which cortical representations are modeled and selected as a function of their use for the organism. Practicing the violin for several hours a day since early childhood may therefore substantially alter a young musician's neuronal networks, their extension, and perhaps even their macroscopic morphology. This is considered the most likely explanation for the expansion of somato-sensory cortex in string players, because the younger the age at which the instrument was played, the greater the effect. Similar radical experience-dependent alterations in cortical topography have been repeatedly observed in the sensory cortex of monkeys. Modern neuroscience thus completely overturns Gall's hypothesis.

Phrenologists considered the cortical surface allocated to a given function as an innate parameter that ultimately determined our level of competence. Quite the contrary, neuroscientists now think that the time and effort one dedicates to a domain modulates the extent of its representation in the cortex.

A decade ago, new studies of Einstein's brain, aroused the attention of the media. Most anatomical measures of that mythical organ, which is preserved in a jar of formaldehyde, were disappointing: The inspired founding father of modern physics seemed to be equipped with a very unexceptional encephalon. Its weight, for instance, was only about 1,200 grams, which is not much even for an old man. However, in 1985, two researchers reported an above-average density of glial cells in a posterior region of the brain called the angular gyrus or Brodmann's area 39, which belongs to the inferior parietal lobule. This area, as we shall see later on, plays a critical role in the mental manipulation of numerical quantities. Hence it was perhaps not unreasonable that its cellular organization should distinguish Einstein from average humans. Had the biological cause of Einstein's excellence finally been exposed?

In fact, this research is plagued by the same ambiguities as the studies of musicians' cortical topography. Even granting that Einstein's cellular density exceeded the normal variability between individuals, which is not yet proved, how can one separate causes from consequences? Einstein may have been endowed from birth with a phenomenal number of inferior parietal cells, predisposing him to learn mathematics. But in the current state of our knowledge, the opposite seems equally plausible: The constant use of this cerebral region may have deeply modified its neuronal organization. Ironically enough, the biological determinants of relativity theory, if any, are thus forever lost in this chicken-and-egg conundrum. Who said that all was relative?

Is Mathematical Talent a Biological Gift?

One argument that has often been exploited to validate the search for the genetic bases of mathematical talent derives from the correlation between the mathematical achievements of siblings, especially between homozygotic twins. Homozygotic twins, who have the same genotype, often seem to exhibit similar levels of performance in mathematics. Heterozygotic twins, who share only half of their genes, appear to be more variable: Occasionally one soars in mathematics while the other stays at a mediocre level. By comparing achievement across many pairs of homozygotic and heterozygotic twins, a measure of "heritability" can be computed. According to studies conducted in the 1960s by Steven Vandenberg, heritability in arithmetic would amount to about 50%—implying that about half the variance in arithmetical performance is due to genetic differences among individuals.

This interpretation, however, remains hotly contested. Indeed, the twins method is at the mercy of many trivial influences. For instance, studies have shown that homozygotic twins receive identical education, in the same classroom, with the same teacher, more often than heterozygotic twins. The fact that they are similarly talented may thus be due to the shared features of their education rather than to their genes. Another potential confound: In their mother's uterus, close to 70% of homozygotic twins share a single placenta or a single set of membranes. This, of course, is never the case for heterozygotic twins, who are born from two separate ova. Thus, the comparable biochemical composition of the uterine environment may perhaps impose common regularities on the developing brains of homozygotic twins. Finally, even if the genetic heritability of mathematical talent were proved, the twins method provides no indication of the genes involved. These could very well have no direct relation to mathematics. To take an extreme example, suppose that a gene influences body size. It could have a negative influence on mathematical abilities simply because its bearers play basketball more often, and their mathematics education suffers!

In the search for the biological bases of mathematical talent, another intriguing though ambiguous cue is provided by differences between men and women. High-level mathematics are almost exclusively a masculine realm. Of the forty-one calculating prodigies described by Steven Smith in his well-documented book on great mental calculators, only three are female. In the United States, Camilla Benbow and her colleagues have administered a test initially designed for teenagers, the Scholastic Aptitude Test for Mathematics (SAT-M), to a large group of 12-years-olds. The average grade is usually around 500 points. For every girl who already exceeds this score in her twelfth year, two boys do. This ratio reaches 4:1 when the grade is raised to 600 points, and to 13:1 beyond 700 (Figure 6.4). Hence the proportion of males increases dramatically as one considers increasingly bright populations of mathematical students. This advantage for males is observed in all countries, from China to Belgium. Men's supremacy in mathematics is a worldwide phenomenon.

The importance of this phenomenon for the general population must be qualified, however. Only the mathematical elite is almost exclusively made up of men. In the population as a whole, men's supremacy is weaker. The impact of gender on a psychological test is measured statistically by dividing the mean difference between men and women by the dispersion of the scores within each gender. In adolescents, this value typically does not exceed one-half, meaning that the distributions of male and female scores overlap considerably: One-third of

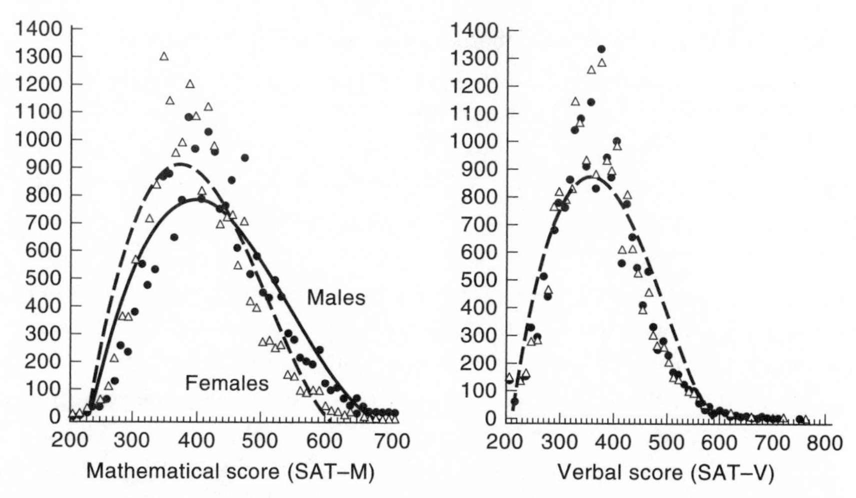

Figure 6.4. In Camilla Benbow's sample of talented seventh-grade students, standard aptitude tests reveal a small but consistent advantage for males over females in mathematics. Verbal scores, in contrast, are distributed identically for males and for females. (Reprinted from Benbow 1988 by permission of the publisher; copyright © 1988 by Cambridge University Press.)

the men fall below the average female score, or conversely, one-third of the women fall above the average male score. The male advantage also varies with the content of the tests. In mathematical problem solving, men clearly take the lead, but in mental calculation, women rank first by a narrow margin. Finally, while a discrepancy between boys and girls emerges from preschool on, no systematic advantage seems detectable before schooling starts. Babies' precocious abilities for arithmetic, in particular, are no more prevalent in males than in females.

In spite of these qualifications, the male hegemony in high-level mathematics raises important issues. Mathematics serves as a filter at several critical stages in our educational systems, and each time, more boys make it than girls. In the end, our society leaves women few opportunities to acquire top-level training in mathematics, physics, or engineering. Sociologists, neurobiologists, and politicians alike would like to know whether this distribution of educational resources justly reflects the natural talents of each gender, or whether it merely serves to perpetuate the biases of our male-governed society.

No doubt, many psychological and sociological factors disfavor women in mathematics. Surveys have shown that, on average, women show greater anxiety than men in mathematics courses; they are less confident in their capacities; they view mathematics as a typically masculine activity that will be of little use in their professional careers; and their parents, especially their fathers, share this feeling. Of course, these stereotypes aggregate into a self-fulfilling prophecy. Young women's lack of enthusiasm for mathematics and their conviction that they will never shine in this domain contribute to their neglect of mathematics courses and hence their lower level of competence.

Very similar stereotypes account for the discrepancies in mathematical achievement according to social class. I am convinced that the prejudices that our societies convey about mathematics are largely responsible for the gap that separates the mathematical scores of men and women, as well as those of rich and poor—a gap that could partially be filled by political and social changes in attitudes toward mathematics. In China, for instance, the most gifted female teenagers obtain mathematical scores that exceed not only those of American female teenagers, but also those of American *male* teenagers—a clear proof that the difference between men and women is small compared to the impact of educational strategies. A recent meta-analysis of dozens of publications suggests that the average gap between American men and women has been reduced by one-half during a thirty-year period, an evolution that parallels the concomitant improvement in the female's status over the same period.

This being said, do biological gender differences play any role in the remaining gap? Though no clear neurobiological or genetic determinants of the male advantage in mathematics have been found yet, a bundle of convergent clues fuels a growing suspicion that biological variables do contribute to mathematical talent, however remotely. In a population of children exceptionally gifted in mathematics, one finds thirteen boys for one girl. Compared to an unselected group of boys and girls, gifted children are also twice as likely to suffer allergies, four times as likely to be myopic, and twice as likely to be left-handed. More than 50% of these budding mathematicians are either left-handers, or ambidextrous, or right-handers with left-handed siblings. Finally, 60% of them are first-born children. Obviously, the archetype of the scholar as a single child, gauche, sickly, and wearing glasses is not totally unfounded!

One might perhaps explain away the association of myopia with mathematical talent by appealing to some attitudinal cause—maybe short-sighted children delve into mathematics books more willingly because they are poor at, say, baseball. A similar argument might be proposed for birth order: Perhaps first-borns receive a subtly different education that somehow encourages mathematical thinking. But allergies and handedness do not easily lend themselves to such a "soft" explanation. Furthermore, there are conclusive, though admittedly more extreme cases in which mathematical capacities are clearly affected by a sex-related neurogenetic anomaly. For instance, a majority of calculating prodigies of the "idiot savant" kind suffer from autism, a neurological disease that strikes boys four times more often than girls. Indeed, autistic symptoms are associated with genetic anomalies of the X chromosome, such as the fragile X syndrome.

Conversely, Turner's syndrome is a genetic disease that affects only women and is linked to a missing X chromosome. As it turns out, in addition to certain physical malformations, women with Turner's syndrome suffer from a profound and specific cognitive deficit in mathematics and in the mental representation of space, even though their IQ may be at a normal level. Their handicap is caused in part by an abnormally feeble secretion of sex hormones due to an atrophy of the ovaries. Indeed, early hormonal treatment is known to improve their mathematical and spatial performance.

We still do not have a satisfactory explanation for these mysterious links between gender, the X chromosome, hormones, handedness, allergies, birth order, and mathematics. All we can do today is paint an impressionist picture of some of the more plausible causal chains—which some scientists have dubbed "just so stories"! According to neuropsychologist Norman Geschwind and his colleagues, exposure to an elevated level of testosterone during gestation might simultaneously affect the immune system and the differentiation of the cerebral hemispheres. Testosterone may slow down the development of the left hemisphere. One can imagine that the likelihood of being left-handed should then increase, as should the ability to manipulate mental representations of space, a function that is more dependent on right-hemispheric processing. This refined sense of space, in turn, would ease the manipulation of mathematical concepts. Because testosterone is a male hormone, this putative cascade of effects could have stronger consequences for males than females. Not implausibly either, it may also be under the partial genetic control of the X chromosome, which may account for the heritability of mathematical and spatial dispositions.

Among the bundles of clues that gravitate around this still fuzzy scenario are these: Androgens are known to directly influence the organization of the developing brain; alterations of the processing of space and mathematics have been demonstrated in subjects exposed to an abnormal level of sex hormones during development, as well as in females at various points in the menstrual cycle; in rats, the spatial abilities of hormonally treated females exceed those of untreated females and catch up with those of untreated males; and finally, the concentration of sex hormones in the womb is higher during the first pregnancy (remember that the majority of mathematical prodigies are first-born). Shaped in this variable hormonal bath, the male brain is probably organized slightly differently from the female brain. Neuronal circuits may be subtly altered in a manner that remains largely unknown so far but which may explain men's slightly swifter motility in abstract mathematical spaces.

It is frustrating to be unable, given the current state of knowledge, to go beyond theoretical fuzziness and to exhibit a simple, determinist account of mathematical talent. But it would surely be naive to expect direct links from genes to genius. The gap is so wide that it can only be filled by a multiplicity of twisted causal chains. Genius emerges from an improbable confluence of multiple factors—genetic, hormonal, familial, and educational. Biology and environment are intertwined in an unbreakable chain of causes and effects, annihilating all hopes of predicting talent through biology or of giving birth to a baby Einstein by crossbreeding two Nobel Prize winners.

When Passion Produces Talent

The limits of a biological account of talent are nowhere more evident than in the case of "idiots savants" with a minuscule island of genius in an ocean of incompetence. Consider the case of Dave, a fourteen-year-old boy who has been studied by Michael Howe and Julia Smith. In an instant, Dave can give the day of the week corresponding to any past or future date. But his IQ does not reach 50, he reads at the level of a six-year-old, and he hardly speaks. Moreover, unlike Michael, whom I described earlier in this chapter, Dave knows close to nothing about mathematics. He is even totally unable to multiply. What biological parameter could possibly have given Dave both a gift for "calendrology" and an aversion to reading and calculation? How could the brain be predisposed to acquire the Gregorian calendar, which has existed in its present form only since 1582? Dave's gift, if there is one, must reside in some generic parameter, such as memory or powers of concentration. To explain the narrowness of his talent, one must obviously appeal to learning. Neither genes nor hormones can instill innate knowledge about the month of December.

It turns out that Dave spends hours at a time scrutinizing the kitchen calendar and drawing it from memory, in part because playing with other children is beyond his social competence. Dave suffers from severe autism. Like a Robinson Crusoe lost in an affective desert, his only companions in solitude are called Friday or January. Suppose that he dedicates three hours a day to calendars (surely an underestimate). In ten years, his training would amount to ten thousand hours of extreme concentration—an enormous duration that may explain both his deep understanding of the calendar and the considerable gaps in his knowledge of all other domains.

From calendar to mental calculation, a similar obsessive concentration characterizes all calculating prodigies, past or present. Why should anyone dedicate all his energy to such a narrow field? Among the great mental calculators, perhaps we should distinguish three main categories: the professionals, the idle, and the mentally deficient. The first are mathematicians in full possession of their mental powers, whose profession requires an in-depth knowledge of arithmetic. For them, calculation can become second nature. Gauss, by his own account, often found himself counting his steps without any conscious intention. As for Alexander Aitken, another brilliant mathematician, he claimed that calculations were set off automatically in his mind: "If I go for a walk and if a motor car passes and it has the registration number 731, I cannot but observe that it is 17 times 43." Not infrequently, as in Gauss's case, such mathematicians lose part of their calculation abilities as they move on to more abstract spheres of the mathematical universe.

In the second category, the idle, I would place calculators whose profession is so dull that they delve in calculation as a pastime. A typical example: Jacques Inaudi and Henry Mondeux, both shepherds, who reinvented much of arithmetic in their lonesome pastures. Both never ceased to count: not only their sheep, but also pebbles, their steps, the time spent balancing on a stool.

Finally, the third category, the mentally deficient, consists of mentally retarded people such as Dave or Michael, who live in an autistic world, and whose passion for numbers or calendars is pathological and symptomatic of their lack of interest for human relations. Jedediah Buxton, an eighteenth-century English calculating prodigy, was most probably autistic. Alfred Binet thus describes Buxton's first night at the theater where *Richard III* was playing:

> He was later asked if the performance had pleased him: he had only seen in it an occasion to calculate; during the dances, he had focused his attention on the number of steps: they numbered 5,202; he had also counted the number of words that the actors had pronounced: this number was 12,445 . . . and all this was found to be exact.

Whatever its motivation, could such an infusion of numbers, year after year, suffice to explain the blossoming of an extraordinary talent for calculation? Could anyone, with sufficient training, turn into a calculating prodigy, or does it take a special, biological "gift"? To tease apart nature from nurture, a few researchers have tried to turn average students into calculating or memory prodigies through intensive training. Their results prove that passion breeds talent. K. Anders Ericsson, for instance, has shown that a hundred hours of training suffice to expand one's digit span to at least twenty digits—eighty digits in one particularly persevering subject. Another psychologist, J. J. Staszewski, has taught a handful of students several strategies for fast calculation. After 300 hours of training spread over two or three years, their calculation speed quadrupled: They took only about thirty seconds to compute mentally 59,451×86.

These learning experiments are in line with the intuitions of the great calculators themselves, who declare that they have to practice daily or else see their talent decline. According to Binet, for instance, "Having dedicated one month to studying books, [Inaudi] saw that he was losing much of his mental powers. His mental calculation abilities only remain stable thanks to ceaseless training."

Alfred Binet also reports a comparison of Jacques Inaudi's calculating speed with that of professional cashiers at the *Bon Marché* in Paris. Prior to automated cash registers, cashier was a respected profession. Genuine human calculators spent eight to ten hours a day, six days a week, adding up purchases and multiplying lengths of linen by the price per meter. Although most were hired between the age of fifteen and eighteen with no particular aptitude for arithmetic, they quickly became lightning calculators. Binet found that they were no slower than Inaudi. Indeed, one of them took only four seconds to compute 638×823, clearly better than Inaudi's six seconds. The sheer extent of his memory, however, enabled Inaudi to win the race in more complex calculations.

The case of the *Bon Marché* cashiers illustrates the absence of any sharp demarcation between professionals whose talent derives from intense training, and geniuses who supposedly owe their feats to an innate gift. Indeed, until recently, the Center for Nuclear Research in Geneva employed Wim Klein for his arithmetic powers; and Zacharias Dase, in the nineteenth century, contributed greatly to mathematics by establishing a table of natural logarithms for numbers 1 through 1,005,000 and by factoring all numbers between 7 and 8 million.

Today, society no longer values mental calculation. Great show-business human calculators are hard to come by. Thus, the professionals of centuries gone by appear all the more prodigious. Nowadays, in the West at least, whoever forced a child to calculate several hours a day would expose himself to a lawsuit —though our society condones the dedication of the same amount of time to piano or chess playing. Oriental societies do not share our value scales. In Japan, it is a well-accepted practice to send children to evening arithmetic courses where they learn the secrets of the "mental abacus." At the age of ten, the most enthusiastic of them can apparently exceed the performance of our Occidental calculating prodigies.

Ordinary Parameters for Extraordinary Calculators

A talent for calculation thus seems to arise more from precocious training, often accompanied by an exceptional or even pathological capacity to concentrate on the narrow domain of numbers, than from an innate gift. This conclusion fits in with the thinking of two of the past centuries' greatest geniuses: Thomas Edison, for whom "genius is 1 percent inspiration and 99 percent perspiration," and the French naturalist Buffon, who confessed—with counterfeit humility?—that "genius is but a greater aptitude for patience."

Supporting this thesis, psychometric studies have not detected any major modification in the fundamental parameters of cerebral functioning of lightning calculators. Outside of their specialty, these prodigies' information-processing speed turns out to be average or slower. Consider Shakuntala Devi, a female Indian calculator with astounding speed. (The Guinness Book of World Records grants her the ability to multiply two thirteen-digit numbers in thirty seconds, although this may be exaggerated.) The psychometrician Arthur Jensen—who has often championed the biological determinism of intelligence in the past—invited her to his laboratory in order to measure her performance on some classical tests. Jensen's article hardly conceals his disappointment: There was nothing exceptional in the time it took this arithmetic genius to detect a light flash or to select one motor action out of eight. Devi's performance in a so-called "intelligence" test, Raven's progressive matrices, did not depart much from average. And when she had to search for a visual target or look for a number in memory, she was abnormally *slow*. To borrow a computer science metaphor, Devi's calculation feats were obviously not due to a global speed-up of her internal clock: Only her arithmetic processor ran with lightning speed.

In the preceding chapter, we saw that one can predict with remarkable precision the time a normal subject will require to do a multiplication. The more elementary the operations needed and the larger the digits concerned, the slower the calculation. In this respect too, calculating prodigies are no different from the average person. A century ago, Binet timed Inaudi while he was solving multiplication problems. Here are some of his results:

	calculation time in seconds	**number of operations**
3×7	0.6	1
63×58	2.0	4
638×823	6.4	9
7,286×5,397	21	16
58,927×61,408	40	25
729,856×297,143	240	36

The column on the right shows how many elementary operations are needed in the traditional multiplication algorithm. This quantity predicts Inaudi's calculation time rather well, with the exception of the most complex multiplication problems, which are disproportionately slow because of the greater memory load. It would be remarkable if Inaudi had been able to multiply two three-digit numbers in barely more time than two single digits. This would indicate that he was using a radically different algorithm, perhaps allowing for the execution of multiple operations in parallel. But this is not the case for Inaudi nor for any other arithmetic genius that I know of. Great calculators struggle with great calculations like the rest of us.

One final characteristic may bespeak an inborn talent: the extraordinary memory exhibited by most lightning calculators. For Binet, this issue was beyond discussion: "In my opinion, memory is the calculating prodigy's essential characteristic. By his memory, he is inimitable and infinitely superior to the rest of humanity."

Binet distinguished two kinds of prodigies, the visual calculators who memorize a mental image of written numbers and calculations, and the auditory ones, like Inaudi, who claim that they remember numbers by hearing them recited in their head. Perhaps one should also add a third category, the "tactile" calculators, since at least one blind lightning calculator, Louis Fleury, maintained that he manipulated numbers mentally as though he were holding some cubarithms, the tactile numerical symbols used by the blind. Regardless of its modality, however, great calculators' memory span is often no less than astounding. Inaudi, for instance, could repeat 36 random digits without error after having heard and repeated them only once. At the end of his daily exhibitions, he never failed to repeat in full the 300-some digits that the public had dictated to him throughout the show.

Undeniably, Inaudi's memory span reached astounding heights, but does this imply that it was innate? Aside from countless anecdotes whose reliability is often questionable, we know little about the childhoods of these prodigies. As yet, nothing proves that they possessed amazing memory abilities at an early age. It seems to me equally plausible that their fantastic memory is the result of years of training as well as their great familiarity with numbers.

Steven Smith, who has carefully studied the lives of dozens of calculating prodigies, reaches the same conclusion: "Mental calculators, no less than other mortals, are subject to short-term memory limitations. Where they differ is in their ability to treat groups of digits as single items in memory."

Memory span, indeed, is not an invariant biological parameter such as blood group that can be measured independently of all cultural factors. It varies considerably with the meaning of the items to be stored. I can easily remember a fifteen-word sentence in French, my first language, because its meaning helps. In Chinese, however, which I do not understand, my memory span drops to about seven syllables. Likewise, perhaps the reason why great calculators manage to store vast amounts of digits is that numbers are almost their mother tongue. There is hardly a combination of digits that does not make sense to them. In

Hardy's memory, the taxi license plate 1729 was probably registered as four independent digits because it looked like any random number. For Ramanujan, however, 1729 was a childhood friend, a familiar character that occupied only one cell in his memory. In general, I think, the extreme familiarity that calculating prodigies have with digits suffices to explain their huge memory span without having to postulate a hypothetical biological gift for number memory.

Recipes for Lightning Calculation

In order to definitely shake off the myth of a "born calculator," I must however explain what algorithms great calculators actually use. Unless I do so, the mental multiplication of 5,498 by 912 or the immediate recognition that 781 is 11×71 will always be enshrouded in a spell of mystery. Most of us, indeed, have not the slightest idea of how to solve such problems mentally. In fact several expedients radically simplify even the most insurmountable-looking arithmetical puzzles.

So how can one compute mentally the product of two multidigit numbers? Scott Flansburg, who became known as "the human calculator," makes no secret of it: His exploits are entirely based on simple recipes that anyone can learn and which he unveils in a recent bestseller. Like all other calculators, he uses calculation algorithms similar to those taught at school. However, the order in which he performs each operation is carefully optimized. For addition, he recommends computation from left to right. For multiplication, he always computes the most significant digits of the result first. Each subproduct is immediately added to the running total, thus avoiding memorization of several long intermediate results. These diverse strategies are headed toward a single goal—minimizing memory load—and they succeed because only a single provisional estimate of the result must be stored and refined step after step.

More rarely, some calculators memorize all or part of the multiplication table for all possible pairs of two-digit numbers. This allows them to multiply by groups of two digits as if they were one. Finally, all calculators possess a huge repertoire of shortcuts based on simple algebraic tricks. To give just one example, the product of 37×39 is immediately identified as $38^2 - 1$ using the formula $(n+1)(n-1) = n^2 - 1$; 38^2 itself equals $36 \times 40 + 4$ since $n^2 = (n-2)(n+2) + 2^2$. One needs only to retrieve from memory the product of 36×4, which any experienced calculator recognizes as $12^2 = 144$, to which one adjoins the digit 3 $(4-1)$, to conclude that 37 times 39 is 1,443! With a little training, applying this method becomes as fast as a reflex.

In brief, great calculators obviously do not rely on any "magical" arithmetic methods. Like us, they rely heavily on stored tables of multiplication facts whose only original features are their extent and, occasionally, their nonverbal format (since some calculators, such as Michael, do not appear to have acquired any language). Like us, they execute their calculations serially, digit after digit, thus explaining Binet's response time measurements. Like us, finally, they quickly select the best means of reaching the result in minimal time from the multiple strategies at their disposal. In this respect, only the number of strategies that they master differentiates them from the six-year-old who already spontaneously simplifies 8+5 into (8+2)+3.

What about more complex arithmetical abilities, though? A single glance is enough for Shakuntala Devi to notice that the seventh root of 170,859,375 is 15 (which means that this number is 15 to the seventh power, or 15×15×15×15×15×15×15). The extraction of roots of integers belongs in the classical repertoire of professional calculators. Naive spectators are always awed by what they consider as a particularly difficult feat, especially for high-order roots. In fact, however, easy shortcuts can dramatically reduce the calculations. For instance, the rightmost digit directly informs us of the corresponding digit of the result. When a number ends with 5, so does its root. In the case of fifth roots, the starting number and its root always end with the same digit. In all other cases, there is a correspondence which is easily learned and which gets even simpler if one considers the last two digits instead of just one. The first digits of the result, on the other hand, can often be found by trial and error using simple approximations. For instance the seventh root of 170,859,375 can only be 15 because 25, the next candidate ending with 5, would obviously yield much too large a number once raised to the seventh power. In brief, extracting the roots of integers, which appears at first sight as a superhuman performance, can be reduced to the careful application of simple recipes.

The ability to rapidly factorize numbers and to identify prime numbers is a more impressive feat. Remember Michael, the autistic man who promptly recognized that 389 is a prime number and that 387 can be decomposed into 9×43? The twins whom Oliver Sacks described were even stranger. Their pastime, it was claimed, consisted of taking turns and exchanging increasingly larger primes up to six, eight, ten, or even twenty digits long!

While this ability seems truly amazing and is still far from being fully understood, several tentative explanations may be proposed. First of all, contrary to a widespread notion, the concept of a prime number is not the pinnacle of mathematical abstraction. Primality is a very concrete notion that merely indicates whether a collection of objects can be divided into several equal groups. Twelve is not prime because it can be divided into three groups of 4 or two groups of 6. Thirteen is a prime because no such grouping is possible. Hence, prime numbers are so common that children manipulate them unknowingly when they try to organize square blocks into a rectangle—they quickly find that it can be done with twelve blocks, but not with thirteen. No wonder, then, that a retarded young man like Michael, with an uncanny passion for arithmetic, can spontaneously discover some of their properties.

Finding out whether a number is prime remains a difficult mathematical problem. Yet the role of memory should not be neglected. There are only 168 prime numbers under one thousand and 9,592 prime numbers under a hundred thousand. Once memorized, they can serve to compute the remaining primes up to ten billion using an obvious algorithm called the sieve of Eratosthenes. Finally, simple recipes known to any schoolboy, such as casting out the 9s, make it easy to determine whether a number is divisible by 2, 3, 4, 5, 6, 8, 9, or 11. Such elementary tricks are apparently all that Michael was using, since he often erred with numbers that looked prime but were in fact the product of factors that exceeded his sagacity (for instance, $391 = 17 \times 23$). What of the twins? Unfortunately, no details are available about the precise numbers they were exchanging or about their potential errors. So we will never know if the method they employed was any more precise than Michael's.

Researchers also often claim that some calculating prodigies can evaluate an exact number of objects at a glance. Binet, for instance, asserted that one could drop a fistful of marbles before Zacharias Dase and that he immediately reported their exact number. Unfortunately, I do not know of any serious psychological study on this purported phenomenon. There have been no measurements of response times, which are the only way to assess whether a person is counting or is really perceiving large numbers "instantaneously." My feeling is that great calculators' enumeration powers do not differ from ours. Confronted with a collection of marbles, their visual system, like ours, rapidly parses it into small groups of one, two, three, or four marbles. Their apparent speed may come from their ability to add all these numbers in a flash, while we are at best reduced to counting by twos.

Finally, many prodigies develop a special ability for calendar calculation. Can this also be attributed to simple strategies? Several well-known algorithms allow one to compute the day of the week for any past or future date. The simpler of them require only a few additions and divisions, and professional calculators no doubt rely on such formulas. However, this explanation does not fit autistic children who become calendrical prodigies. Most of them have never had access to a perpetual calendar. One blind boy's talent developed even though he never had access to a Braille calendar! Furthermore, some prodigies, such as Dave, are unable to perform even the simplest of calculations. What, then, are the tricks through which they compute the days of the week?

By timing the responses of several autistic prodigies, Beate Hermelin and Neil

O'Connor have discovered that their response time is generally proportional to the distance that separates the requested date from the present. This suggests that most of these "human calendar calculators" use a very simple method: Starting with a recent date, they proceed by degrees and progressively extrapolate to the nearby weeks, months, or years. Many regularities facilitate this extrapolation process: The calendar repeats itself every 28 years; weeks shift by one day for each regular year and two days for leap years; March and November always start on the same day, and so on. Most idiot savants use such knowledge to jump directly from, say, March 1996 to November 1968. Thus they can instantly retrieve from memory the requested page of the calendar, from which they merely have to read the appropriate date.

How can such an algorithm, however simple, be invented and faultlessly executed by an idiot savant whose IQ does not exceed 50? Dennis Norris, a Cambridge researcher, has developed an interesting computer simulation of the acquisition of calendrical knowledge in a neural network. His simulated network comprises several hierarchical neuronal assemblies that successively receive inputs coding for the day, the month, and the year of a random date between 1950 and 1999. At the output, seven units code for the seven days of the week. Initially, the network does not know what day it should associate with a given date. As it receives more and more examples—Monday, April 22, 1996, or Sunday, February 3, 1969, and so on—it progressively adjusts the weight of its simulated synapses in order to adapt to the difficult task of predicting on what day each date will fall. After several thousand trials, not only does it retain these examples, but it also responds correctly to more than 90% of the novel dates that it has never learned. Hence, the final network exhibits good knowledge of the mathematical function that relates dates and days of the week—knowledge that is only implicit, since its synapses ignore anything about subtraction and addition, or even the number of days in a year or the existence of leap years.

According to Norris, the nervous system is equipped with learning algorithms far superior to those he used in his simulation. It thus appears entirely plausible that an autistic child, even one severely retarded, who spends years studying the calendar, may extract a mechanical, automated, and unconscious knowledge of it by mere induction on many examples.

Talent and Mathematical Invention

In the final analysis, where does mathematical talent come from? Throughout this chapter, every track that we have explored has led us to a plausible source.

Genes probably play a part. But by themselves they cannot supply the blueprint of a phrenological "bump" for mathematics. At best, together with several other biological factors, perhaps including precocious exposure to sex hormones, genes may minimally bias cerebral organization to aid the acquisition of numerical and spatial representations. Biological factors, however, do not weigh much when compared to the power of learning, fueled by a passion for numbers. Great calculators are so passionate about arithmetic that many prefer the company of numbers to that of fellow humans. Whoever dedicates that much time to numbers must succeed equally well in increasing memory and in discovering efficient calculation algorithms.

If only one lesson might be drawn from this analysis of talent, it would be that high-level mathematics departs radically from its popular portrayal as a dryly rational discipline, dominated by sheer deductive power, on which emotions have no bearing. Quite the contrary, the most potent of human emotions—love, hope, pain, or dispair—hold sway over the relationship these mathematicians entertain with their number friends. When there is a passion for mathematics, talent does not lag very far behind. If, conversely, a child develops math anxiety, this phobia can prevent even the simplest of mathematical concepts from falling into place.

My survey of mathematical talent has given equal footing to Ramanujan and Michael, Gauss and Dave, the genius and the idiot savant. Yet can one really compare the giants that extend the frontiers of mathematics and the autistic prodigies that shine only because of the striking contrast between their mathematical abilities and their profound mental retardation? My choice is justified by the many characteristics that geniuses and calculating prodigies share—from their passion for mathematics to their vision of a landscape populated by numbers. In my opinion, it would be unfair to deny Inaudi or Mondeux the name of "genius" under the pretext that they merely rediscovered well-known mathematical results. When a shepherd, alone in his pasture, rediscovers Pythagoras's theorem, his talent is no less than that of his renowned predecessor to whose work he was never exposed.

In this chapter I have deliberately avoided dwelling on the psychological and neurobiological preconditions that underlie mathematical creativity. The flash of invention is so brief that it can hardly be studied scientifically. At best one can speculate, as did Jean-Pierre Changeux and Alain Connes, that scientific discovery involves the more or less random association of old ideas, followed by a selection based on the harmony and adequacy of a newly formed combination. Paul Valéry said, "It takes two persons to invent: one forms the combinations, the other chooses and recognizes what is desired or relevant among the set of products of the first." Augustine likewise noted that *cogito* means "to shake together," while *intelligo* means "to select among."

Jacques Hadamard, in his major investigation of invention in mathematics, distinguishes stages of preparation, incubation, illumination, and verification. Incubation consists of an unconscious search through fragments of demonstrations or original combinations of ideas. In support of this central idea, Hadamard quoted Henri Poincaré: "Most striking at first is this appearance of sudden illumination, a manifest sign of long, unconscious prior work. The role of this unconscious work in mathematical invention appears to me incontestable."

Some day we will perhaps understand the cerebral bases of this "cognitive unconscious." The spontaneous activity of neuronal circuits below the threshold of consciousness, the unleashing of automatic calculation mechanisms during sleep—these must have measurable physiological traces that we can hope to assess with modern brain imaging tools. At present, however, we can only heed the question that Hadamard asked already half a century ago: "Will it ever happen that mathematicians will know enough about the physiology of the brain and neurophysiologists enough of mathematical discovery for efficient cooperation to be possible?"

Indeed, we will now look into brain physiology—not in the hope of uncovering the biological bases of creativity, which would be a Utopian dream given the current state of our knowledge; but at least to try to explain how the rudimentary paraphernalia of neurons, synapses, and receptor molecules incorporate into the brain's circuits the routine of calculation and the meanings of numbers.

Part III

Of Neurons and Numbers

Chapter 7

Losing Number Sense

> The true idea of the human mind is to consider it as a system of different perceptions or different existences, which are linked together by the relation of cause and effect, and mutually produce, destroy, influence, and modify each other. . . . In this respect, I cannot compare the soul more properly to anything than to a republic or commonwealth, in which the several members are united by the reciprocal ties of government and subordination.
>
> David Hume, *A treatise of human nature*

Is it possible that one could forget what 3 minus 1 means, while remaining able to read and write four-digit numerals? Can you imagine being able to multiply digits that appear on your right, but not those that appear on your left? Is it possible, finally, for someone with normal vision to fail on written additions as simple as 2+2, while easily solving the same problems when they are read aloud?

Strange as they may seem, such phenomena are routinely observed in neurology. Cerebral lesions of various origins can have a devastating and sometimes surprisingly specific impact on arithmetic abilities. Everybody knows that a lesion in the motor areas of the brain can cause paralysis to one side of the body only. By the same mechanism, brain damage confined to the cerebral areas involved in language or number processing can alter only a very narrow domain of competence. The lesion seems to have few repercussions until the patient is asked to subtract or to read an unusual word, and then a profound deficit is unveiled.

As early as 1769, the French philosopher Denis Diderot anticipated the specificity of neurological impairment. In *D'Alembert's Dream*, he made this premonitory statement:

> According to your principles, it seems to me that with a series of purely mechanical operations, I could reduce the greatest genius in the world to a mass of unorganized flesh . . . [The operation] would consist in depriving the original bundle of some of its threads and shuffling up the rest. . . . Example: take away from Newton the two auditory threads, and he loses all sense of sound; the olfactory ones, and he has no sense of smell; the optic ones, and he has no notion of colors; the taste threads, and he cannot distinguish flavors. The others I destroy or jumble, and so much for the organization of the man's brain, memory, judgment, desires, aversions, passions, willpower, consciousness of self.

Cerebral lesions are indeed devastating events that can destroy the brightest minds. Yet to neuroscientists these "experiments of nature" also offer a unique glimpse into the workings of the normal human brain. Cognitive neuropsychology is the scientific discipline that takes advantage of data from patients with brain lesions to gather knowledge about the cerebral networks that serve cognitive functions. The neuropsychologist's touchstone is *dissociation*, or the fact that after cerebral damage, one domain of competence becomes inaccessible while another remains largely intact. When two mental abilities are thus dissociated, one may often safely infer that they involve partially distinct neuronal networks. The first ability is deteriorated because it normally requires the contribution of a cerebral area that has been damaged and is now unable to perform. The second remains intact because it rests on cerebral networks that have been spared by the lesion. Of course, neuropsychologists must beware that more trivial explanations for a dissociation exist. For instance, one task might simply be easier than the other, or the patient might have relearned one ability but not the other after the lesion occurred. When care is taken to reject such alternative accounts, cognitive neuropsychology supports remarkable inferences about cerebral organization.

Let us consider a concrete example. Michael McCloskey, Alfonso Caramazza, and their colleagues have described two patients with severe difficulties in reading Arabic numerals. The first patient, known to us only by his initials H. Y., occasionally misreads number 1 as "two" or 12 as "seventeen." A careful study of his errors shows that while H. Y. often replaces one numeral with another, he never errs in the decomposition of a number into hundreds, tens, and units. For instance he reads 681 as "six hundred *fifty*-one"—the structure of the string is correct except for the substitution of *fifty* for *eighty*. Conversely, the second patient, J. E., never takes 1 for "two" or 12 for "seventeen," but he misreads 7,900 as "seven thousand ninety" or 270 as "twenty thousand seventy." Unlike H. Y., J. E. does not substitute one number word for another. Instead, the whole grammatical structure of the numeral is wrong. He recognizes individual digits, but they wander from the hundreds to the decades or the thousands column.

Patients H. Y. and J. E. together realize a *double dissociation*. Schematically, the grammatical structure of numerals is intact in H. Y. and deteriorated in J. E., while the selection of individual words is intact in J. E. and deficient in H. Y. The very existence of two such patients suggests that some of the cerebral regions engaged in reading Arabic numerals aloud contribute more heavily to number grammar, while others are more concerned with accessing a mental lexicon for individual number words. If the lesion was small enough—unfortunately, an infrequent event with vascular lesions—its location could even provide valuable indications as to exactly where in the brain these areas lie.

In interpreting such observations, one must of course beware of falling back into phrenology. If patient J. E. errs in the grammar of numerals, this does not mean that his lesion knocked out "the grammar area." Broad cognitive faculties such as "grammar" are complex and integrated functions that likely imply the concerted orchestration of several distributed areas of the brain. Most likely, J. E.'s lesion affected a highly specialized elementary neuronal process essential to the production of a grammatical sequence of number words, but not for the selection of its component words.

The extreme modularity of the human brain stands out as the main lesson to be gathered from studies of cerebral pathology. Each small region of the cortex appears to be dedicated to a specific function and may thus be viewed as a mental "module" specialized in processing data from a distinct source. Cerebral lesions and the bizarre dissociation patterns they provoke provide us with a unique source of information on the organization of these modules. Thanks to dozens of handicapped patients such as H. Y. and J. E. who generously agreed to participate in scientific experiments, our knowledge of the cerebral areas involved in number processing is now much deeper than it was barely ten years ago. To be sure, the exact circuits used in complex arithmetic operations still escape us. Yet an increasingly refined map of the cerebral pathways for numerical information is slowly taking shape. Even the rudimentary knowledge we currently have of the neurology of number processing already has considerable bearing on our understanding of the relations between mathematics and the brain.

Mr. N, the Approximate Man

As Mr. N enters the examination room on a morning in September 1989, the devastating effects of his cerebral lesion are obvious. His right arm is in a sling, and his crippled right hand betrays a severe motor handicap. Mr. N speaks slowly, with effort. Occasionally, he searches with growing irritation for a very common word. He cannot read a single word, and he fails to understand such moderately complicated commands as "Place the pen on the card, then put it back in its original location."

Mr. N was once married and is the father of two daughters. He held a position of responsibility as a sales representative in a major firm, and he doubtless was proficient in arithmetic. We know little about the circumstances in which his world shattered. He apparently suffered a bad fall at home, perhaps due to sudden brain hemorrhaging. Upon his arrival at the hospital, he suffered from an enormous hematoma, and an emergency operation was performed. These dramatic events left him with a vast lesion of the posterior half of the left hemisphere. Three years later, his language and motor control handicaps are still so devastating that he cannot lead an independent life and lives with his elderly parents.

My colleague Dr. Laurent Cohen invited me to meet Mr. N because he suffers from exceptionally severe *acalculia,* the neurologist's technical term for a deficit in number processing. We ask him to calculate two plus two. After pondering for a few seconds, he answers "three." He easily recites the rote numerical series 1, 2, 3, 4 . . . and 2, 4, 6, 8 . . . , but when we ask him to count 9, 8, 7, 6 . . . or 1, 3, 5, 7 . . . , he fails completely. He also fails to read the digit 5 when I flash it before his eyes.

Given this distressing clinical picture, it would tempting to conclude that Mr. N's arithmetic abilities are as good as gone, as is most of his competence for language. Yet several observations contradict this hypothesis. First is Mr. N's strange reading behavior. When I make him see the digit 5 for an extended period of time, he manages to tell me that it is a digit, not a letter. Then he starts counting on his fingers—"one, two, three, four, five, it's a five!" Obviously he must still recognize the shape of the digit 5 in order to count up to the appropriate numeral. But why can't he then immediately utter it? When I ask him how old his daughter is, he behaves similarly. Unable to access the word "seven" instantly, he covertly counts up to this numeral. He appears to know right from the start what quantities he wishes to express, but reciting the number series seems to be his only means of retrieving the corresponding word.

In passing, I notice a similar phenomenon when Mr. N attempts to read words aloud. He often gropes around for the appropriate meaning without finding the right word. While unable to read the handwritten word *ham*, he manages to tell me, "It's some kind of meat." The word *smoke* is equally unreadable but evokes a sense of "having a fire, burning something." He confidently reads the word *school* as "classroom." The direct pathway that enables any of us to move straight from the sight of digit 5 to its pronunciation "five," or from the letters h–a–m to the sound "ham," seems to have vanished from Mr. N's mind. Nevertheless, in one way or another the meaning of these printed characters is not totally lost for him, and he clumsily attempts to express it using circumlocutions.

Following up this lead, I next show Mr. N a pair of digits, 8 and 7. It would take him several seconds to "read" them by counting on his fingers. Yet in a twinkling he readily points out that 8 is the larger digit. Much the same occurs with two-digit numerals, which he experiences no difficulty in classifying as larger or smaller than 55. Mr. N obviously remembers the quantity represented by each Arabic numeral. His only errors occur when the quantities are similar, like 53 and 55. It is as if he only knows their approximate magnitude. He also manages to place two-digit numerals at their approximate location on a vertical line labeled 1 at the bottom and 100 at the top, which is presented to him as a thermometer. His responses, however, are far from being digitally accurate. He places 10 at the lower quarter, while 75 lands much too close to 100. Operating finer classifications is impossible to him. Deciding whether a number is odd or even, in particular, widely exceeds his capacities.

In experiment after experiment, a striking regularity emerges: Though Mr. N has lost his exact calculation abilities, he can still approximate. Every task that calls only for an approximate perception of numerical quantities poses no difficulty for him. On the one hand, he easily judges whether a certain quantity is roughly appropriate to a concrete situation—for example, nine children are in a school: Is this too few, just right, or too many? On the other hand, he has obviously lost all precise memory for numbers. He judges that a year comprises "about 350 days" and an hour "about fifty minutes." According to him, a year has five seasons, a quarter of an hour is "ten minutes," January has "fifteen or twenty days," and a dozen eggs make for about "six or ten eggs"—responses that are both clearly false and yet not that far from the truth. Even his immediate memory has not been spared. When I flash the digits 6, 7, and 8 at him, a second later he cannot remember if he has seen a 5 or a 9. Yet he is quite confident that neither 3 nor 1 were among the initial set because he quickly realizes that these numbers represent too small a quantity.

The dissociation between exact and approximate knowledge is nowhere more apparent than in addition. Mr. N does not know how to add 2+2. His random responses 3, 4, or 5 testify to his profound acalculia. Yet he never offers a result as absurd as 9. Likewise, when presented with a slightly wrong addition, such as 5+7=11, he judges it to be correct more than half of the time, thus confirming that he cannot compute its exact result. Yet he can rapidly reject with total confidence and complete success a grossly false answer such as 5+7=19. He apparently still knows its approximate results and he quickly detects that the proposed quantity, 19, departs from it by a lot. Interestingly, the larger a quantity, the fuzzier it seems to be in Mr. N's mind. Thus he rejects 4+5=3 but accepts 14+15=23. Multiplication problems, however, seem to exceed the scope of his approximation abilities. He answers them in a seemingly random fashion, even accepting as correct an operation as absurd as 3×3=96.

In a nutshell, Mr. N suffers from a peculiar affliction: He is unable to go beyond approximation. His arithmetic life is confined to a strange, fuzzy universe in which numbers fail to refer to precise quantities and have only approximate meanings. His torments refute the cliché of the unfailing precision of mathematics, so elegantly expressed by the French writer Stendhal: "I used to love, and still love, mathematics for themselves as a domain that does not admit *hypocrisy* and *vagueness*, my two pet aversions."

With all due respect to Stendhal, vagueness is an integral part of mathematics —so central, in fact, that one may lose all exact knowledge of numbers and yet maintain a "pure intuition" of numerical quantities. Wittgenstein was closer to the truth when he maliciously observed that 2+2=5 is a reasonable error. But if an individual asserts that 2+2 make 97, then this cannot just be a mistake: this person must be operating with a logic totally different from our own.

In earlier chapters, I drew a distinction between two categories of arithmetic skills: the elementary quantitative abilities that we share with organisms devoid of language, such as rats, apes, and human babies; and the advanced arithmetic abilities that rest on symbolic notations of numbers and on the strenuous acquisition of exact calculation algorithms. Mr. N's case suggests that those two categories rely on partially separate cerebral systems. One can be abolished while the other remains intact.

It would obviously be absurd and reductive to equate patient N's performance with that of Sheba, Sarah Boysen's gifted chimpanzee whom I described in the first chapter. For all his handicaps, Mr. N remains a full-fledged *Homo sapiens*. In arithmetic, however, his cerebral lesion has thrown him back to a rudimentary level of competence. Like Sheba, Mr. N can go from a numerical symbol to the corresponding quantity—although his repertoire of symbols is evidently much larger than the chimp's. Like her, he is also able to select the larger of two quantities and to compute an approximate addition. That these operations remain accessible to an aphasic and acalculic patient with a drastically impaired left hemisphere confirms that they do not depend much on linguistic abilities. Exact calculation, on the other hand, calls for the integrity of neuronal circuits specific to the human species and localized at least in part in the left hemisphere. This is why Mr. N, with his extended left-hemispheric lesion, can neither read numbers aloud, nor multiply them, nor judge whether they are odd or even.

A Clear-Cut Deficit

Mr. N's case does not allow for very strong conclusions about the cerebral localization of numerical approximation. Given the extent of his lesion in the left hemisphere, his residual abilities may well rest on intact areas of the right hemisphere. However, the possibility remains that part of his left hemisphere has remained functional enough to allow for number comparison and approximation, if not exact calculation.

Other neurological pathologies are better suited to pinpointing the arithmetic abilities of each hemisphere. The corpus callosum is a massive bundle of nerve fibers that connects the two hemispheres and that serves as the main pathway for communicating information between them. Occasionally, this bundle can be disconnected. Sometimes it is partially interrupted by a focal brain lesion. More frequently, it is purposely severed surgically in an effort to control severe epilepsy in patients not amenable to any other form of treatment. In either case the result is a human being with a cortex divided in two, or a *split-brain* patient. The two cerebral hemispheres remain in full working order, but it is now practically impossible for them to exchange any information.

In everyday life, these patients appear deceptively sound in body and mind. Their behavior seems entirely normal—except for very rare episodes where their left hand undoes what their right hand is doing. A simple neurological examination, however, suffices to reveal clearcut deficits. If the patients close their eyes and a familiar object is placed in their left hand, they are unable to name it, though they can demonstrate its use through gestures. Likewise, if a picture is flashed within their left visual field, they swear that they haven't seen anything, but their left hand manages to select the appropriate picture among many others.

This odd behavior can easily be accounted for. The major neuronal projection pathways that connect the external sense organs to primary sensory cortices are crossed, so that a tactile or visual stimulation from the left side is initially processed by the sensory areas of the right hemisphere. Thus, when an object is placed in the left hand, the right hemisphere is fully informed of the identity of the stimulus and can retrieve its shape and function. Yet in the absence of the corpus callosum, this information cannot be transmitted to the left hemisphere. In particular, the cerebral areas that control language production, whose lateralization to the left hemisphere has been known since the work of Broca in the last century, are given no indication of what the right hemisphere feels or sees. The left-hemispheric language network thus denies having seen anything. If it is compelled to provide an answer, it selects a response at random or borrows it from previous trials. That was the case in my testing of a patient who, while blindfolded, had just named a hammer placed in her right hand. When I placed a corkscrew in her left hand, she immediately said "another hammer"—all the while her left hand mimicked unscrewing a bottle.

Patients with a severed corpus callosum are a gold mine for neuropsychologists because they allow for a systematic assessment of the cognitive abilities available in each hemisphere. Suppose that one asks a split-brain patient to multiply a digit by 2 and point at the appropriate result placed among several other numbers. By presenting the digit visually either to the right or to the left of the patient's gaze, and by flashing it so briefly that it is gone before the eyes have had time to move, one can ensure that the input remains confined to a single hemisphere. Using this trick, it becomes possible to assess whether either hemisphere can identify numbers, multiply them by 2, or allow the patient to point toward a given number.

Let us start with the simplest operation: identifying digits. Flash two digits on a screen and ask a split-brain patient whether they are identical or different. When one digit appears to the right and the other to the left, even this simple same-different judgment is not feasible. The patient responds at random, sometimes deciding that 2 and 2 are different, and sometimes that 2 and 7 are identical. The severance of interhemispheric connections makes comparing the digits on the left and right impossible. This is so even if each hemisphere, on its own, can identify them. Indeed, when the two digits appear in the same visual field, either both on the right or both on the left, the patient responds with almost perfect accuracy.

The two hemispheres do not stop at recognizing digit shapes. They can also interpret them as referring to a certain quantity. To prove this, one can present a digit together with a set of dots rather than a pair of digits. When both the digit and the dot pattern appear in the same visual field, the patient easily determines whether they match. Thus, each hemisphere knows that 3 and ∴ represent one and the same number.

Both hemispheres also appreciate the ordinal relation between numbers. Whether a digit is presented to the right or to the left, split-brain patients can quickly decide whether it is smaller or larger than some reference number. And when a pair of digits is flashed, they can point toward the larger (or toward the smaller). Comparison merely seems to be a bit slower and less accurate in the right hemisphere than in the left, but the difference is small. Hence, each hemisphere appears to host a representation of numerical quantities and a procedure for comparing them.

But this similarity of the two hemispheres vanishes when one tackles the issue of language and mental calculation. These functions are the left hemisphere's indisputable privilege. Using the same experimental procedures as just described, the right hemisphere appears unable to identify written numerals. Its visual abilities include the recognition of simple shapes such as the digit 6, but not of alphabetical stimuli such as *six*. In most people, the right hemisphere is also mute: It cannot produce most words aloud. Thus, if one flashes the digit 6 on the left-hand side of a computer screen, the vast majority of split-brain patients behave exactly as Mr. N would: They cannot name the digit, although they can indicate with the left hand that this number is larger than 5.

Some particularly ingenious patients manage to circumvent their right hemisphere's incapacity to produce speech. For instance, Michael Gazzaniga and Steven Hillyard have studied a patient called L. B. who, after several seconds, managed to name digits presented to his right hemisphere. Unlike a normal person, his naming time increased linearly with digit size: it took him two seconds to name digit 2, but almost five seconds to name digit 8. Like Mr. N, L. B. appeared to recite the number sequence slowly and covertly until he had reached a numeral that "stuck out"—those were his own words—and which he then uttered aloud. Nobody knows exactly how the right hemisphere managed to signal that the number it had seen had been reached. It might have been some kind of hand movement, a contraction of the face, or some other cueing artifice that split-brain patients often devise for themselves. Anyhow, the very fact that the patient resorted to counting in order to name digits presented in the left visual field indicates that his right hemisphere was devoid of normal speech production abilities.

The right hemisphere is also ignorant of mental arithmetic. When an Arabic digit is presented in the right visual field and therefore contacts the left hemisphere, the patient experiences no apparent difficulty adding 4 to it, subtracting 2 from it, multiplying it by 3, or dividing it by 2. Such calculations, however simple, are strictly impossible when the digit appears on the left side and is therefore processed by the right hemisphere. This profound calculation deficit persists even when the patient is asked to point toward the result rather than say it.

Though the right hemisphere is worthless for exact calculation, can it nevertheless approximate? To assess this possibility, my colleague Laurent Cohen and I asked a patient with partially disconnected hemispheres to verify visually presented addition problems. Even when the operation was as obviously wrong as 2+2=9, when it was perceived by the right hemisphere the patient seemed to respond randomly and judged it to be correct on about half of the trials. During one series of trials, however, she suddenly had a run of fifteen correct responses out of sixteen. The probability that such an event could occur by chance is less than 1 in 4,000. I therefore believe that her right hemisphere could estimate simple additions, but managed to express this competence only during this single block of sixteen trials. Indeed, it is not enough for the right hemisphere to possess a certain ability; it must also understand the experimenter's instructions and be given

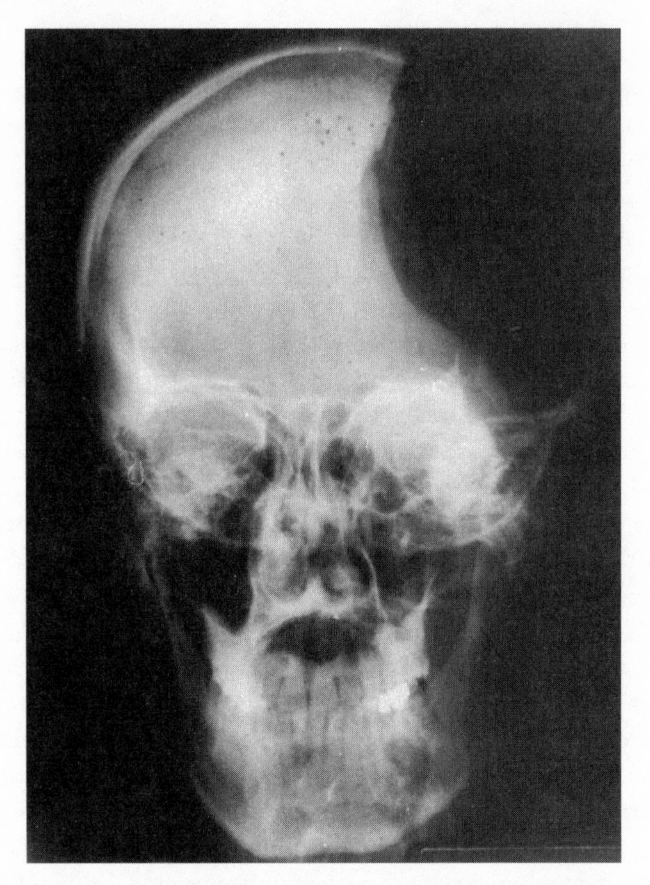

Figure 7.1. Despite the loss of his left hemisphere during combat in Vietnam, patient J.S. can still identify and compare Arabic numerals. Exact calculation, however, poses him extreme difficulties. (Reprinted from Grafman et al. 1989 by permission of the publisher.)

a chance to respond before the left hemisphere takes over.

Jordan Grafman and his colleagues have studied another patient who brings further support to the hypothesis that the right hemisphere is good only at very elementary calculations. A young American soldier, J. S., lost most of his left skull and underlying cortex at the age of 22 during combat in Vietnam (Figure 7.1). Somehow, J. S. survived the many surgical operations, repetitive infections, and severe epilepsy that ensued. He now lives a semi-independent life with a single right hemisphere (in the left hemisphere, only the occipital lobe is spared). As can be expected, J. S. is profoundly impaired in spoken language comprehension and production. He can neither read nor write, and he cannot name any object—deficits that coincide exactly with the known limitations of the isolated right hemisphere in patients with a severed corpus callosum. His results on tests of number processing are also in keeping with those of other split-brain studies. J. S. recognizes Arabic numerals and knows how to compare them and estimate the numerosity of a set of objects. He occasionally reads aloud a few digits and some two-digit numbers. He can solve only about half of the single-digit addition and subtraction problems set to him. Multiplication, division, and multidigit calculation constitute an insurmountable challenge for him.

A Champion in Numerical Non-sense

The split-brain patients we have seen, together with patient J. S., indicate that although only the left hemisphere can perform exact calculation, both the left and the right hemispheres incorporate representations of numerical quantities. Can the brain areas implicated in this quantitative representation be localized? Is the mental number line associated with a specific cerebral circuit that occupies a precise cortical location? And what would our mental life be like if a brain lesion made us lose our number sense? To answer these questions, I turn to patients with smaller lesions that affect a more specific piece of brain circuitry.

When the famous writer Eugene Ionesco was working on his masterpiece *The Lesson*, he probably had few pretensions other than a love of humor and nonsense. Yet in this play, he unknowingly sketched a remarkably realistic portrait of an acalculic patient devoid of any quantitative intuition:

PROFESSOR: Let us arithmetize a little now. . . . How much are one and one?

PUPIL: One and one make two.

PROFESSOR: *marveling at the Pupil's knowledge*: Oh, but that's very good. You appear to me to be well along in your studies. You should easily achieve the total doctorate, miss. . . . Let's push on: how much are two and one?

PUPIL: Three.

PROFESSOR: Three and one?

PUPIL: Four.

PROFESSOR: Four and one?

PUPIL: Five. . . .

PROFESSOR: Magnificent. You are magnificent. You are exquisite. I congratulate you warmly, miss. There's scarcely any point in going on. At addition you are a past master. Now, let's look at subtraction. Tell me, if you are not exhausted, how many are four minus three?

PUPIL: Four minus three? . . . Four minus three?

PROFESSOR: Yes. I mean to say: subtract three from four.

PUPIL: That makes . . . seven?

PROFESSOR: I am sorry but I'm obliged to contradict you. Four minus three does not make seven. You are confused: four plus three makes seven, four minus three does not make seven . . . This is not addition anymore, we must subtract now.

PUPIL, *trying to understand*: Yes . . . yes . . .

PROFESSOR: Four minus three makes . . . How many? . . . How many? . . .

PUPIL: Four?

PROFESSOR: No, miss, that's not it.

PUPIL: Three, then.

PROFESSOR: Not that either, miss . . . Pardon, I'm sorry . . . I ought to say, that's not it . . . excuse me.

PUPIL: Four minus three . . . Four minus three . . . Four minus three? . . . But now doesn't that make ten? . . .

PROFESSOR: Count then, if you will, please.

PUPIL: One . . . two . . . and after two, comes three . . . then four . . .

PROFESSOR: Stop there, miss. Which number is larger? Three or four?

PUPIL: Uh . . . Three or four? Which is the larger? The larger of three or four? In what sense larger?

PROFESSOR: Some numbers are smaller and others are larger. In the larger numbers there are more units than in the small. . . .

PUPIL: Excuse me, Professor . . . What do you mean by the larger number? Is it the one that is not so small as the other?

PROFESSOR: That's it, miss, perfect. You have understood me very well.

PUPIL: Then, it is four.

PROFESSOR: What is four—larger or smaller than three?

PUPIL: Smaller . . . no, larger.

PROFESSOR: Excellent answer. How many units are there between three and four? . . . Or between four and three, if you prefer?

PUPIL: There aren't any units, Professor, between three and four. Four comes immediately after three; there is nothing at all between three and four! . . .

PROFESSOR: Look here. Here are three matches. And here is another one, that makes four. Now watch carefully—we have four matches, I take one away, now how many are left?

PUPIL: Five. If three and one make four, four and one make five.*

---

*Source: E. Ionesco, *The Lesson,* (translated by Donald M. Allen). English translation copyright ©1958 by Grove Press Inc. Used by permission of Grove/Atlantic Inc.

Did Ionesco ever visit a neurology clinic? *The Lesson*'s pupil is not an imaginary character, but someone whom I have met in person. For several hours, I attempted to teach arithmetic to Mr. M, a sixty-eight-year-old acalculic patient with a lesion of the inferior parietal cortex (Figure 7.2). Like Ionesco's pupil, this person could still solve simple additions, but he was totally unable to subtract and had trouble determining the larger of two digits. Ionesco's dialogue rings so true that it could almost be a verbatim transcription of my surrealistic conversations with Mr. M. In the professor's lines I recognize my own clumsy attempts at teaching Mr. M elementary arithmetic; my disproportionate encouragement when he succeeded; and my barely concealed discouragement in front of his recurring failures. In the pupil's words, I can almost hear my patient's confusion as he tried, with an unfailing willingness, to answer questions that he no longer understood. Even the play's subtitle—"a comical drama"—fits to a T Mr. M's unfortunate predicament, a genuine case of numerical nonsense.

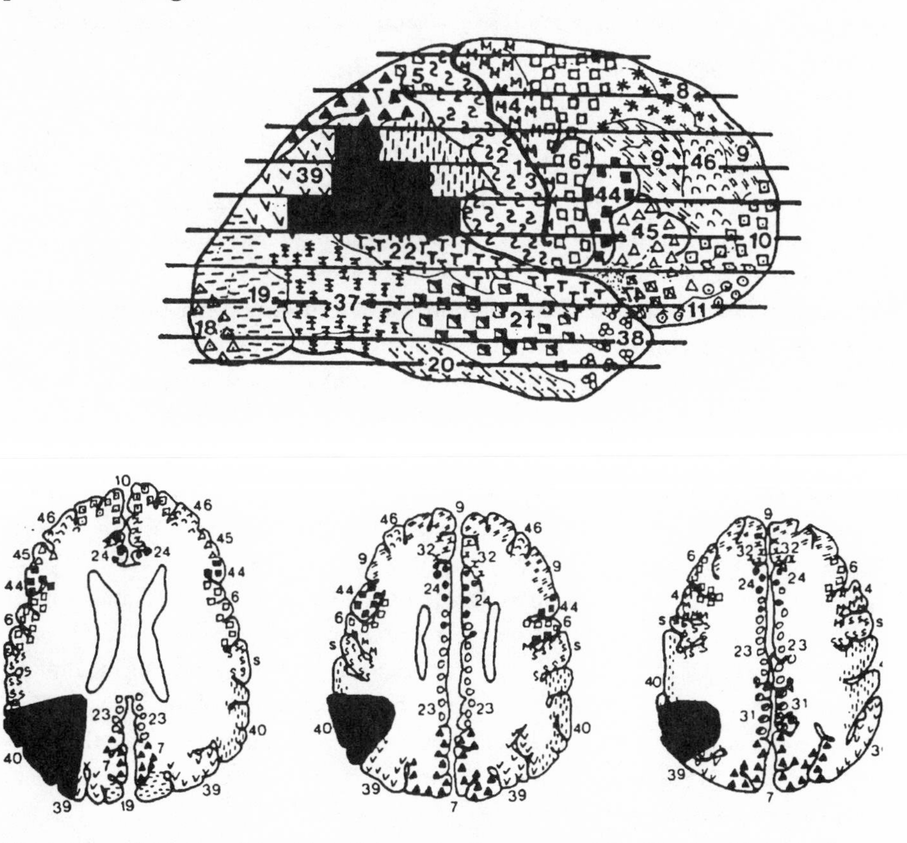

Figure 7.2. *This lesion of the right inferior parietal cortex caused Mr. M to lose his sense of numerical quantities. (Note that a confusing neurological convention makes the right-hemisphere lesion appear on the left of horizontal sections.) (From Dehaene and Cohen, 1996.)*

Mr. M's impairment is in fact typical of patients who suffer from a selective deficit of the quantitative representation of numbers, the mental number line that gives meaning to Arabic numerals and number words. Mr. M has essentially lost every intuition about arithmetic. This is why he is unable to compute four minus three or even to figure out what this subtraction might mean. Nevertheless, because his other cerebral circuits remain intact, he can still perform routine symbolic calculations, while at the same time failing to understand them.

Let us consider Mr. M's dissociated abilities one at a time. Mr. M speaks quite fluently and can read words and numbers to perfection. He initially suffered from some difficulty with writing, but this handicap has long since receded. His modules for identifying words, both visually and auditorily, and for speaking or writing them must therefore be intact, as are the bundles of connections that connect them. In passing, Mr. M's case forcefully suggests that there are direct pathways in the human brain for transforming numerals from one notation to the other—networks capable of turning 2 into *two* without caring about the meaning of the symbols.

Indeed, Mr. M does not understand the numbers he reads so well. In a task of number comparison that calls for pointing to the larger of two Arabic numerals, he fails once in every six trials. His errors, though relatively infrequent, are gross. For instance, he once maintained, without blinking, that 5 was larger than 6. In a test of number proximity, which consists of deciding which of two numbers is closer to a third, he also fails once in every five trials.

His handicap is most flagrant in subtraction and in number bisection tests. The bisection test consists in deciding which number falls exactly in the middle of a given interval. Mr. M's responses verge on complete nonsense. Between 3 and 5, he places 3, then 2; between 10 and 20, he places 30, only to later correct his answer to 25 with this telling apology: "I do not visualize numbers very well."

A similar confusion reigns over subtraction. He fails to solve about three subtraction problems out of four. And indeed, his mistakes have an eerie resemblance to those of Ionesco's pupil. Two minus one makes two, he affirms. Nine minus eight is seven "because there is one unit." Three minus one "makes four, no there is one unit, a modification of one unit makes three, doesn't it?" For six minus three, he writes down nine, but comments in a rare moment of lucidity: "I am adding when I should be subtracting. Subtracting means taking away; adding means summing up." This knowledge, however, is nothing more than a theoretical veneer. Mr. M has lost all sense of the structure of integers and of which operations are required to move from one quantity to another.

In *The Lesson*, the pupil who cannot subtract three from four suddenly turns out to be a calculating prodigy:

PROFESSOR: How much, for example, are three billion seven hundred fifty-five million nine hundred ninety-eight thousand two hundred fifty one, multiplied by five billion one hundred sixty-two million three hundred and three thousand five hundred and eight?

PUPIL, *very quickly*: That makes nineteen quintillion three hundred ninety quadrillion two trillion eight hundred forty-four billion two hundred nineteen million one hundred sixty-four thousand five hundred and eight. . . .

PROFESSOR, *stupefied*: But how did you know that, if you don't know the principles of arithmetical reasoning?

PUPIL: It's easy. Not being able to rely on my reasoning, I've memorized all the products of all possible multiplications.

All things considered, Mr. M exhibits a similar, though necessarily less spectacular, dissociation. He who confidently asserts that 3–2=2 still knows most of the multiplication table by heart. His rote verbal memory is intact and allows him to blurt out "three times nine is twenty-seven" like an automaton, without understanding what he is saying. He also appeals to this intact memory in order to solve more than half of the one-digit addition problems posed to him. He fails, however, whenever the result of an addition goes beyond ten. The strategy used by most adults, which consists in decomposing, say, 8+5 into (8+2)+3, is out of his reach. Mr. M's arithmetic knowledge starts to dwindle at the point where his rote memory stops. His inferior parietal lesion prevents him from having recourse to the number sense when his memory fails.

Inferior Parietal Cortex and the Number Sense

The inferior parietal area, which is the seat of Mr. M.'s lesion, remains a *terra incognita* of the human brain. This cortical area, particularly its posterior convolution called the "angular gyrus" or "Brodmann's area 39," plays a crucial role in the mental representation of numbers as quantities. It might well be the depository of the "number sense" to which this book is dedicated, an intuition of quantities present ever since the dawn of humankind. Anatomically, it lies in what neuroscientists used to call the high-level plurimodal association cortex. The neurologist Norman Geschwind called it an "association area of association areas." Its neural connections indeed place it at the convergence of highly processed data streams stemming from vision, audition, and touch—an ideal location for arithmetic because the number concept applies equally well to all sensory modalities.

Almost sixty years have elapsed since the German neurologist J. Gerstmann first described the tetrad of deficits that a lesion of the left inferior parietal region can cause: acalculia, needless to say, but also difficulties in writing, in representing the fingers of the hand, and in distinguishing left from right. Immediately after his vascular accident, Mr. M exhibited all these deficits. There was, however, one additional complication: Mr. M's lesion was located in the *right* hemisphere. We believe that this patient, who was strongly left-handed, fell into a minority of people whose brain is organized in a mirror image of its normal architecture and whose right hemisphere is involved in language processing rather than the left. But the loss of quantitative number sense can also be found in more classical patients whose Gerstmann's syndrome stems from a *left* inferior parietal lesion.

What is the relationship between numbers, writing, fingers, and space? This issue is a matter of considerable debate. The tetrad of deficits called Gerstmann's syndrome may not mean much. It could merely reflect the clustering of an odd assortment of independent cerebral modules in the same cortical neighborhood. Indeed, researchers have observed for decades that the four elements making up the syndrome, though frequently found together, can also be dissociated. Some relatively uncommon patients show isolated acalculia with no apparent impairment in distinguishing their fingers, or vice versa. Hence, the inferior parietal region is probably subdivided into microregions highly specialized for numbers, for writing, for space, and for the fingers.

It is nevertheless tempting to look for a deeper explanation for this grouping within the same general brain region. After all, as we saw in previous chapters, the association between numbers and space is indisputably close. In Chapter 1, we saw that numerosity can be extracted from a spatial representation of sets of items, provided this map specifies the presence of objects regardless of their size and identity. In Chapter 3, the mental representation of integers on a left-to-right oriented number line turned out to play a central role in numerical intuition. In Chapter 6, finally, tight relationships were found between mathematical talent and spatial abilities. Little wonder, then, if we find that a lesion can simultaneously destroy mental representations of space and of numbers.

My feeling is that the inferior parietal region hosts neural circuitry dedicated to the representation of continuous spatial information, which turns out to be ideally suited to the coding of the number line. Anatomically, this area stands at the top of a pyramid of occipitoparietal areas that construct increasingly abstract maps of the spatial layout of objects in the environment. Number emerges naturally as the most abstract representation of the permanence of objects in space—in fact, we can almost define number as the only parameter that remains constant when one removes object identity and trajectory.

The links between numbers and fingers are also obvious. All children in all cultures learn to count on their fingers. It thus seems plausible that in the course of development, the cortical representations of fingers and of numbers come to occupy neighboring or tightly interrelated cerebral territories. Furthermore, the cerebral representations of numbers and the layout of the hand, even if they are dissociable, obey very similar principles of organization. When Mr. M wiggles his index finger even though I have asked him to move his middle finger, his error seems to be the exact analogue of his inability to visualize the respective locations of numbers 2 and 3 on the number line. From this perspective, which remains highly speculative, body maps, spatial maps, and the number line would all result from a single structural principle governing the connectivity in the inferior parietal cortex.

Seizures Induced by Mathematics

Another enigmatic pathology demonstrates the extent to which the inferior parietal area is specialized for arithmetic. *Epilepsia arithmetices* is a syndrome first reported in 1962 by the neurologists D. Ingvar and G. Nyman. During a routine electroencephalographical examination of an epileptic girl, they discovered that whenever their patient solved arithmetic problems, even very simple ones, her brain waves showed rhythmic discharges. Calculation triggered epileptic fits, while other intellectual activities such as reading had no effect.

Nimal Senanayake, a Sri Lankan physician, paints a fascinating and terrifying portrait of these "seizures induced by thinking":

> A 16-year-old school girl had been experiencing sudden jerky movements of her right arm during the past year, accompanied by transient thought block when studying; in particular, when studying mathematics. During the term test, she began to develop jerks about 30 minutes after starting the mathematics paper. The pen dropped out of her hand and she found it difficult to concentrate. She completed the 1-hour paper with difficulty but during paper 2 the jerks became more pronounced and in 45 minutes she had a grand mal convulsion and lost consciousness. [Following anti-epileptic medication,] there was some improvement but she continued to have occasional jerks during mathematics lessons. About 9 months after the first major seizure she had to sit the main examination. Again, during the mathematics paper, she started to jerk within 15 minutes. She forced herself to continue, but halfway through the paper she had a grand mal convulsion.

More than a dozen similar cases of "arithmetic epilepsy" are now known throughout the world. The victims' electro-encephalogram frequently presents anomalies in the inferior parietal region. Most likely this area houses an incorrectly wired and hyperexcitable network of neurons that, when put to use during arithmetic problem solving, transmits an uncontrollable electrical discharge to other brain areas. That this epileptic focus only breaks out during calculation gives an indication of the extreme specialization of this cerebral area for arithmetic.

The Multiple Meanings of Numbers

Mr. M's case also provides ample proof of the amazing specialization of the inferior parietal area. Though his parietal lesion has devastated his number sense, Mr. M maintains an excellent knowledge of nonnumerical domains. Most strikingly, although he cannot tell which number falls between 3 and 5, the very same bisection task applied to other areas does not give him any difficulty. He knows very well which letter falls between A and C, which day comes between Tuesday and Thursday, which month falls between June and August, and which musical note is found between do and mi. Knowledge of these series is fully intact. Only the series of numbers—the only one that refers to quantity—seems to be affected.

Even with regard to numbers, Mr. M has not lost his wealthy store of "encyclopedic" knowledge. This talented artist, now retired, can still lecture for hours on the events of 1789 or 1815. He has even told me, with a wealth of numerical detail, the history of the *Hôpital de la Salpêtrière* where I test him. Number 5, which he so readily judges to be greater than 6, evokes in him a profusion of mystical references to the "five pillars of Islam." He reminds me that odd numbers, according to the Pythagoreans, were the only ones that found favor in the gods' eyes. And the patient humorously refers me to a whimsical quote by the French humorist Alphonse Allais: "Number 2 rejoices in being so odd." No doubt, then, Mr. M's erudition has survived brain damage, even in regard to dates and the history of numbers and mathematics.

Another dimension of Mr. M's impairment is that it varies according to the abstractness or concreteness of the problems he is asked to solve. The numbers that are manipulated in arithmetic are highly abstract concepts. When solving 8+4, there is no point in wondering whether one is talking about eight apples or eight children. Mr. M's handicap seems confined to this understanding of numbers as abstract magnitudes. His numerical performance improves considerably whenever he finds a concrete referent or mental model to cling to, rather than having to work with numbers in the abstract. For instance, he can still estimate unfamiliar but concrete magnitudes such as the duration of Columbus's trip to the New World, the distance from Marseilles to Paris, or the number of spectators at a major football game. During one examination, he failed to divide 4 by 2 (he mechanically responded, "Four times three is twelve"). Attempting to understand the source of his failure, I placed four marbles in his hand and asked him to share them between two children. He immediately divided this concrete set by grabbing two marbles in each hand without even a shadow of indecision.

Later on I questioned him about his daily schedule and found that he judiciously uses time labels. Mr. M easily explained how he got up at five in the morning and then had two hours of work before breakfast, which was served at seven, and so on. Moving mentally on the concrete line of time was a breeze for him compared to dwelling on the abstract number line. Remarkably, he was able to perform computations with time labels that he was completely unable to perform in the abstract. For instance, he could tell me how much time elapsed, say, between 9 a.m. and 11 a.m.—an operation equivalent to subtraction, which he had so much difficulty with. One peculiarity of the French system of time is that we use both a twelve-hour format and a twenty-four-hour format for time—for instance, we say that 8 p.m. is literally "20 o'clock." Mr. M experienced no difficulty at all converting back and forth between these two formats, although such a conversion is formally equivalent to adding and subtracting 12. As expected, he experienced a bitter setback when I presented him numerically equivalent operations such as 8+12 in the abstract context of an arithmetic test.

These dissociations illustrate how useless it would be to seek *the* brain area for number meaning. Numbers have multiple meanings. Some "random" numbers such as 3,871 refer only to a single concept, the pure quantity that they convey. Many others, however, especially when they are small, evoke a host of other ideas: dates (1492), hours (9:45 p.m.), time constants (365), commercial brands (747), zip codes (90210, 10025), phone numbers (911), physical magnitudes (110/220), mathematical constants (3.14 . . . ; 2.718 . . .), movies (2001), games (21), and even drinking laws (21 again!). The inferior parietal cortex seems to encode only the quantitative meaning of numbers, which is what Mr. M has trouble with. Distinct brain areas must be involved in coding the other meanings.

In Mr. G, another patient with massive damage to the left hemisphere, the contribution of these parallel pathways for number meaning is particularly evident. Mr. G suffers from a major reading deficit. The direct reading pathway that converts written letters or digits into the corresponding sounds is totally disrupted, preventing him from reading most words and numbers. Yet some strings still evoke fragments of meaning:

- 1789: It makes me think of the takeover of the Bastille . . . but what?
- Tomato: It's red . . . one eats it at the beginning of a meal . . .

Sometimes this semantic approach allows him to recover the pronunciation of a word in a very indirect way:

- 504 [a famous model of Peugeot car]: The number of the cars that win . . . it was my first car . . . it begins with a P . . . Peugeot, Renault . . . it's Peugeot . . . 403 [another Peugeot!] . . . no 500 . . . 504!
- Candle: One lights it to light up a room . . . Candle!

On other occasions, conversely, the retrieved meaning leads him astray:

- 1918: the end of World War I . . . 1940
- Giraffe: zebra

Though pure quantities can reasonably be related to inferior parietal cortex, nobody knows yet which cerebral areas take on the other nonquantitative meanings of numbers. Among the many unsolved issues that the cognitive and neural sciences will have to address in the next ten or twenty years, this one certainly stands out: According to what rules does our brain endow a linguistic symbol with meaning?

The Brain's Numerical Information Highways

The meaning of numbers is not the only knowledge that is distributed among several brain regions. Think of all the arithmetic know-how you command: reading and writing numbers, in Arabic or in spelled-out notation; understanding them and producing them aloud; addition, multiplication, subtraction, division—and the list can go on. The study of cerebral lesions suggests that each of these abilities rests on a swarm of highly specialized neuronal networks communicating through multiple parallel pathways. In the human brain, division of labor is not an idle concept. Depending on the task that we plan to accomplish, the numbers that we manipulate go down different "cerebral information highways." A small part of these networks is tentatively schematized in Figure 7.3

Consider reading. Do we use the same neuronal circuits to identify the Arabic digit 5 and the word *five*? Probably not. Visual identification as a whole rests on cerebral areas in the posterior part of both hemispheres, in a region called the

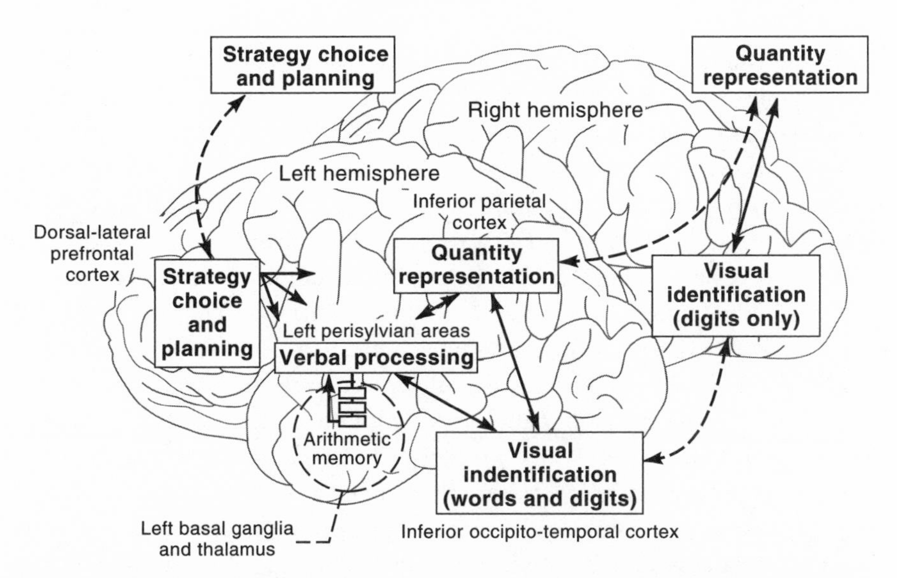

Figure 7.3. A partial and still hypothetical diagram of cerebral areas involved in number processing. Both hemispheres can manipulate Arabic numerals and numerical quantities, but only the left hemisphere has access to a linguistic representation of numerals and to a verbal memory of arithmetic tables. (After Dehaene and Cohen 1995.)

inferior occipito-temporal cortex. However, this region is highly fragmented into specialized subsystems. The study of split-brain patients indicates that the visual system of the left hemisphere recognizes both Arabic numerals and spelled-out words, while that of the right hemisphere recognizes only simple Arabic numerals. Furthermore, even within the posterior left hemisphere, different categories of visual objects—words, Arabic digits, but also faces and objects—seem to be processed by dedicated neuronal pathways. Hence, certain lesions of the left occipito-temporal region impair only the visual identification of words. These patients suffer from a syndrome called "pure alexia" or "alexia without agraphia." *Alexia* means that they cannot read a word (though they understand spoken language perfectly); *without agraphia* means that they can still write words and sentences—though they are totally unable to read their own writing only seconds after having written. Here is a typical transcript of a pure alexic patient attempting to read the word *girl*:

PATIENT: That's 'on' . . . that's 'O, N' . . . 'on' . . . is that what it is? Well there's three letters, like an 'E, B' . . . I don't know what that says . . . I can't see it that well . . . I have to give up, I can't.

EXAMINER: Try to read the letters one by one.

PATIENT: These? It's . . . 'B' . . . 'N' . . . 'I' . . . I don't know.

Though incapable of identifying words, such patients often maintain excellent face and object recognition abilities. Thus, visual identification is not impaired as a whole. Instead, only a subsystem specialized for strings of characters goes awry. Most important for our present purposes, even the identification of Arabic digits is frequently preserved. One of the first diagnosed cases of pure alexia, reported by the French neurologist Jules Déjerine in 1892, involved a man who could not decipher words nor, oddly enough, musical notation, but was still able to read Arabic digits and numerals and even carry out long series of written calculations. In 1973, the American neurologist Samuel Greenblatt described a similar case in which, in addition, the patient still had fully intact visual fields and color vision.

The converse dissociation is also on record. Lisa Cipolotti and her colleagues at the National Hospital in London recently observed a deficit in reading Arabic numerals in a patient who experienced no difficulty reading words. Such cases imply that word and number identification rest on distinct neuronal circuits in the human visual system. Because they lie in neighboring anatomical areas, they frequently deteriorate simultaneously. In some rare cases, however, we can demonstrate that they are in fact distinct and dissociable.

Similar patterns of dissociation are found between writing down numbers and saying them aloud. Patient H. Y., whom I described briefly at the beginning of this chapter, mixed up number words when he had to say them aloud. Yet he experienced no difficulty in writing them in Arabic notation. Thus he might say that "two times five is thirteen," but he always wrote down 2×5=10 correctly. He clearly had preserved a memory for multiplication tables. He failed only when he tried to retrieve the pronunciation of the result. Frank Benson and Martha Denckla similarly described a patient who, when solving 4+5, said *eight* and wrote down 5—yet could still point to the correct result, 9, among several other digits! This patient's cerebral routines for the spoken and written production of numerals were both deteriorated, yet visual identification and calculation remained unaffected.

The extraordinary selectivity of cerebral lesions seems perpetually to catch us off guard. Patrick Verstichel, Laurent Cohen, and I studied a patient who, when trying to speak, emits an incomprehensible jargon ("I margled the tarboneek placidulagofalty stoch . . . "). A careful analysis of errors shows that a specific stage of speech production, which assembles the phonemes making up the pronunciation of words, is irremediably impaired. Yet number words somehow escape this jargon. When the patient tries to say a numeral, say *twenty-two*, he never produces muddled speech like "bendly daw." Like H. Y., however, he occasionally substitutes one number word for another and says *fifty-two* (such whole-word substitutions rarely, if ever, occur with words other than numerals). Thus, even deep down within the stream of cerebral areas for speech production, specialized neuronal circuits deal with the assembling of numerals.

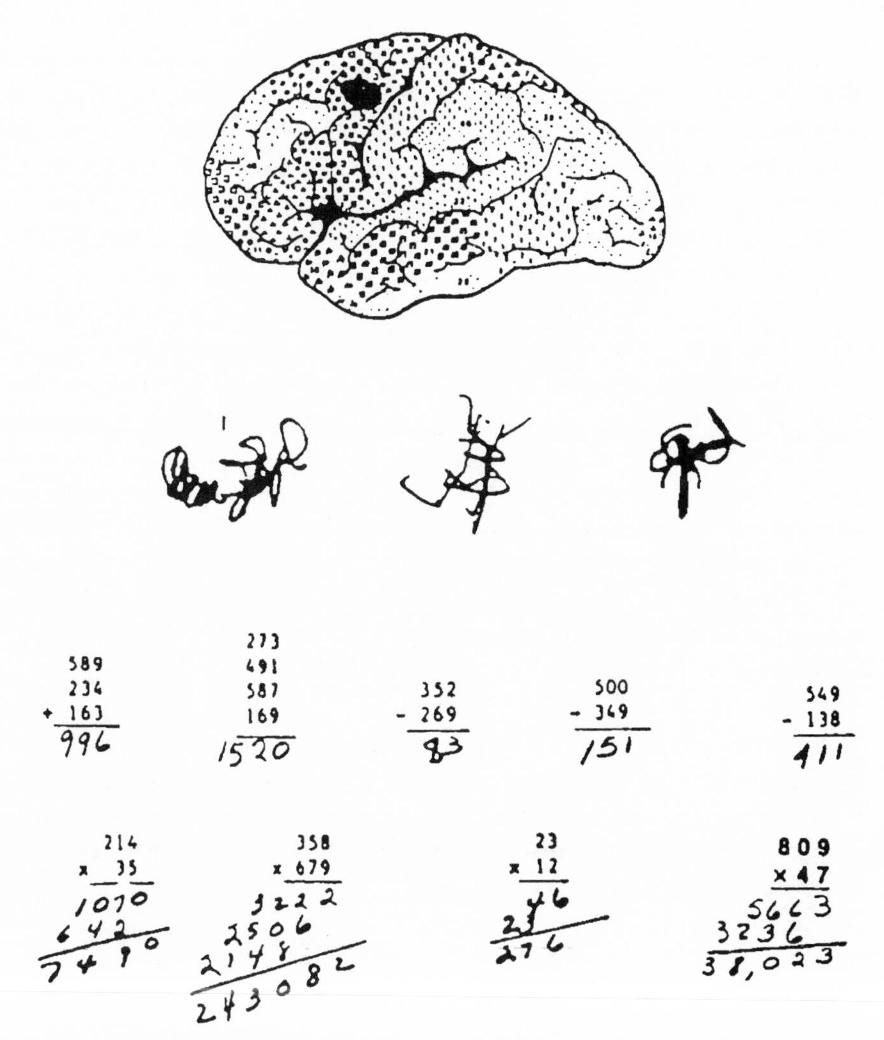

Figure 7.4. *Following a small lesion of the left premotor cortex, this woman became unable to read or write words, but could still read and write Arabic numerals. The scribbles reflect the patient's attempt to write her name, the letters A and B, and the word* dog. *Sample calculations show that her writing of Arabic numerals was fully spared. (Reprinted from Anderson et al. 1990; copyright © 1990 by Oxford University Press.)*

A very similar dissociation is found in writing. Steven Anderson and Antonio and Hannah Damasio have described a patient who had suddenly become unable to read or write after a minuscule lesion destroyed part of her left premotor cortex. When she was asked to write down her name or the word *dog*, all she could produce were illegible scrawls. Yet reading and writing of Arabic numerals remained fully intact. The patient could still solve complex arithmetic problems in the same neat handwriting that she had possessed before the lesion (Figure 7.4).

An inescapable conclusion from this series of analogous cases is that at almost all levels of processing—visual identification, language production, writing—the cerebral areas that handle numerals are partially distinct from those that deal with other words. Many of these areas are not shown on Figure 7.3, for the simple reason that we do not yet know much about their anatomical substrate. But their dissociation following a cerebral lesion proves at least that they do exist.

Let us now talk about calculation. We have already described at length the crucial role of the inferior parietal cortex in the quantitative processing of numbers and in particular in their subtraction. But what about addition and multiplication tables? My colleague Laurent Cohen and I believe that another neuronal circuit may be involved—a cortico-subcortical loop involving the basal ganglia of the left hemisphere. The basal ganglia are neuronal nuclei located below the cortex. They collect information from several cortical regions, process it, and send it back through multiple parallel circuits passing through the thalamus. Although the exact function of these cortico-subcortical loops remains poorly understood, they are involved in the memorization and reproduction of automatic motor sequences, including verbal sequences. Laurent Cohen and I think that one of those circuits is activated during multiplication and automatically blurts out, say, the result "ten" as a complement to the word sequence "two times five." More precisely, the activity of a distributed population of neurons coding for the sentence "two times five" activates neurons within circuits of the basal ganglia that, in turn, light up a population of neurons coding for the word "ten" within cortical language areas. Other verbal automatisms such as proverbs, poems, or prayers may be stored in a similar fashion.

Our speculations are supported by several cases of acalculia stemming from a left subcortical lesion. Damage to the deep neuronal pathways of the left hemisphere, which leaves the cortex intact, occasionally causes arithmetic impairments. I recently examined a patient, Mrs. B, whose left basal ganglia had been damaged. In spite of this lesion, the patient can read numbers and write them to dictation. Her circuits for identifying and producing numbers are fully intact. The subcortical lesion has had a drastic impact on calculation, however. In fact, Mrs. B's memory for arithmetic tables is so severely disorganized that she now makes mistakes even on problems as simple as 2×3 or 4×4.

In sharp contrast with Mr. M, who had lost number sense, Mrs. B still shows an excellent understanding of numerical quantities (her inferior parietal cortex has been fully spared). She can compare two numbers, find which number falls in between them, and even recalculate 2×3 by mentally counting three groups of two objects. She also experiences no difficulty in solving simple subtractions such as 3–1or 8–3. The narrow domain in which Mrs. B is impaired concerns the retrieval of familiar word sequences from rote memory. She can no longer recall strings of words that once were highly familiar, such as "three times nine is twenty-seven" or "two four six eight ten." In a memorable working session, I asked Mrs. B to recite the multiplication table, the alphabet, some prayers, some nursery rhymes, and some poems, and discovered that all of these forms of rote verbal knowledge are impaired. Mrs B experiences profound deficiencies when she recites *Au clair de la lune*, a nursery rhyme that is about as famous in France as *Twinkle twinkle little star* is in the United States. She cannot recite the alphabet beyond A B C D. She also mixes up the words of the Confiteor, the Apostles' Creed, and the Our Father (which she once ended thus: "and do not forgive but may thy kingdom come"). These deficits are all the more striking because Mrs. B is a devout Christian and a recently retired schoolteacher. Thus, she had spent a lifetime reciting these words. Whether multiplication tables, prayers, and nursery rhymes are stored in exactly the same circuits is unclear. But at the very least, they seem to recruit parallel, probably neighboring neuronal networks of the basal ganglia that were destroyed simultaneously by Mrs. B's subcortical lesion.

Up to now, this book has been concerned only with elementary arithmetic. But what about more advanced mathematical abilities, such as algebra? Should we postulate yet other neuronal networks dedicated to them? Recent discoveries by the Austrian neuropsychologist Margarete Hittmair-Delazer seem to suggest so. She has found that acalculic patients do not necessarily lose their knowledge of algebra. One of her patients, like Mrs. B, lost his memory of addition and multiplication tables following a left subcortical lesion. Yet he could still recalculate arithmetic facts by using sophisticated mathematical recipes that indicated an excellent conceptual mastery of arithmetic. For instance, he could still solve 7×8 as $7\times10 - 7\times2$. Another patient, who had a Ph.D. in chemistry, had become acalculic to the point of failing to solve 2×3, $7-3$, $9\div3$, or 5×4. He could nevertheless still execute abstract formal calculations. Judiciously making use of the commutativity, associativity, and distributivity of arithmetic operations, he was able to simplify $\frac{a\times b}{b\times a}$ into 1 or $a\times a\times a$ into a^3, and he recognized that the equation $\frac{d}{c}+a=\frac{d+a}{c+a}$ is generally false. Although this issue has been the matter of very little research to date, these two cases suggest, against all intuition, that the neuronal circuits that hold algebraic knowledge must be largely independent of the networks involved in mental calculation.

Who Orchestrates the Brain's Computations?

The scattering of arithmetic functions in a multitude of cerebral circuits raises a central issue for neuroscience: How are these distributed neuronal networks orchestrated? How do dispersed cerebral regions recognize that they all code for the same number in different formats? Who or what decides to activate such and such circuits, in a precise order, as a function of the required task? How does the unity of consciousness, the feeling that we experience of executing a calculation one step after the other, emerge from the collective functioning of multiple parallel neuronal assemblies, each holding a small fraction of arithmetic knowledge?

Neuroscientists have not reached a definite answer yet. The current theory, however, is that the brain dedicates specific circuits to the coordination of its own networks. These circuits largely rely on areas located in the front of the brain, notably the prefrontal cortex and the anterior cingulate cortex. They contribute to the supervision of novel, nonautomated behaviors—planning, sequential ordering, decision making, and error correction. It has been said that they constitute a kind of "brain within the brain," a "central executive" who autonomously regulates and manages behavior.

Some of these terms are so vague that they barely belong in our scientific vocabulary yet. They sometimes recall the infamous *homunculus*, the little man dear to Tex Avery and Walt Disney who, comfortably seated at the command post of brain, directs the other body organs—but who directs him? Another *homunculus*? For most researchers, these models are but provisional metaphors. They are destined to be heavily revised as the frontal sectors of the brain are progressively divided into well delimited areas, each assuming a restricted and manageable function. Without doubt, no such thing as *the* frontal system exists. Prefrontal areas comprise a multitude of networks specialized for working memory, error detection, or setting a course of action. Their collective behavior ensures the appearance of a supervised coordination of cerebral activity.

Prefrontal areas play a key role in mathematics, including arithmetic. As a rule, a prefrontal lesion does not affect the most elementary operations, but it can yield a specific impairment in executing a series of operations in the appropriate order. Not infrequently, neuropsychologists come across frontal patients who have become unable to use the multiplication algorithm. They add when they should multiply, they do not process digits in the correct order, they forget to carry over when needed, or they mix up intermediate results—often the telltale signs of a basic inability to supervise a sequence of operations.

Prefrontal cortex is especially vital for the on-line maintenance of the intermediate results of a calculation. It provides a "working memory," an internal representational workspace that allows the output of a computation to become the input to another. Thus, an excellent test of frontal lesions consists in asking a patient to subtract 7 successively, starting from 100. Although frontal patients generally get the first subtraction right, they often jumble the following ones or fall prey to some repetitive response pattern such as 100, 93, 83, 73, 63, and so on.

Arithmetic word problems of the type used in elementary schools worldwide also reveal the contribution of prefrontal areas. Frontal patients fail to design a reasoned resolution strategy. Rather, they often impulsively rush to the first calculation that jumps to mind. A typical case was described by the famous Russian neuropsychologist Aleksandr Romanovitch Luria:

> A patient with a lesion of the left frontal lobe was given the problem just stated: "there were 18 books on two shelves, and there were twice as many books on one as on the other. How many books were on each shelf?" Having heard (and repeated) it, the patient immediately carried out the operation 18÷2=9 (corresponding to the portion of the problem "there were 18 books on two shelves"). This was followed by the operation 18×2=36 (corresponding to the portion "there were twice as many on one shelf"). After repetition of the problem and further questioning, the patient carried out the following operations: 36×2=72; 36+18=54, etc. Characteristically, the patient himself is quite satisfied with the result obtained.

Tim Shallice and Margaret Evans have shown that many frontal patients also experience difficulties in "cognitive estimation": They frequently provide absurd answers to simple numerical questions. One patient declared that the highest building in London was between 18,000 and 20,000 feet tall. When his attention was drawn to the fact that this was higher than the 17,000 feet he had previously attributed to the highest mountain in Britain, he merely reduced his estimate of the highest building to 15,000 feet! According to Shallice, such simple but unusual questions simultaneously call for the invention of novel strategies for numerical estimation, and for an evaluation of the plausibility of the retrieved result. Both components—planning and verification—seem to be pivotal functions of the "central executive" to which prefrontal regions make a main contribution.

With my American colleagues Ann Streissguth and Karen Kopera-Frye, I assessed numerical estimation in teenagers whose mothers drank heavily during pregnancy. Intrauterine exposure to alcohol can have dramatic teratogenic effects. Not only does it alter body development (children born of an alcoholic mother have characteristic facial features that confer a family resemblance on them); it also tampers with the laying down of cerebral circuits, causing microcephalia and abnormal neuronal migration patterns in various brain regions, including prefrontal cortex. Indeed, the teenagers we tested, although they could all read and write numbers and perform simple calculations, provided truly nonsensical numerical responses in cognitive estimation tasks. The size of a large kitchen knife? Six feet and a half, said one of them. The duration of a drive from San Francisco to New York? An hour. Curiously, although their numerical answers were often quite wrong, the patients almost always selected appropriate units of measurement. Sometimes they even seemed to know the answers, yet they still selected an inappropriate number. When asked to estimate the height of the tallest tree in the world, one patient correctly reported "redwood," then generously granted it precisely 23 feet and 2 inches!

The prefrontal cortex, so adept at executive functioning, is one of the cerebral regions most unique to humans. Indeed, the emergence of our species was accompanied by a huge increase in the size of frontal areas, to such an extent that they represent about a third of our brain. Their synaptic maturation is particularly slow—evidence shows that prefrontal circuits remain flexible at least up to puberty and probably beyond. The prolonged maturation of prefrontal cortex might explain some of the systematic errors to which all children in certain age groups fall prey. I am thinking in particular of the Piagetian tests that tap into the "nonconservation of number." Why do young children impulsively respond on the basis of the length of a row of objects, even when they are so competent in number processing? The fault may well lie in the immaturity of their frontal cortex, which makes them unable to inhibit a spontaneous but incorrect tendency. An immature "central executive" may also account for class inclusion errors in which children judge that, in a bunch of flowers made up of eight roses and two tulips, there are more roses than flowers. Such "childishness" may well be symptomatic of a lack of supervision of behavior by the prefrontal cortex. And conversely, the frontal region is among the first to feel the effects of cerebral aging. We can recognize several aspects of the frontal syndrome in "normal" aging: inattention, deficiencies in planning, and perseveration of error, with a preservation of daily routine activities.

At the Origins of Cerebral Specialization

Let me now sketch a summarized model of how the human brain incorporates arithmetic. Numerical knowledge is embedded in a panoply of specialized neuronal circuits, or "modules." Some recognize digits, and others translate them into an internal quantity. Still others recover arithmetic facts from memory or prepare the articulatory plan that enables us to say the result aloud. The fundamental characteristic of these neuronal networks is their modularity. They function automatically, in a restricted domain, and with no particular goal in sight. Each of them merely receives information in a certain input format and transforms it into another format.

The computational power of the human brain resides mostly in its ability to connect these elementary circuits into a useful sequence, under the sway of executive brain areas such as the prefrontal cortex and the anterior cingulate. These executive areas are responsible, under conditions that remain to be discovered, for calling the elementary circuits in the appropriate order, managing the flow of intermediate results in working memory, and controlling the accomplishment of calculations by correcting potential errors. The specialization of cerebral areas allows for an efficient division of labor. Their orchestration, under the aegis of the prefrontal cortex, brings about a flexibility that is invaluable for the design and execution of novel arithmetic strategies.

Where might the extreme specialization of several cerebral areas for number processing come from? Since time immemorial, approximate numerical quantities have been represented in the animal and human brains. A "quantitative module," which may include circuits within the inferior parietal cortex, therefore belongs to the genetic envelope of our species. But what should we think of the specialization of occipito-temporal cortex for the visual recognition of digits and letters, or of the implication of the left basal ganglia in multiplication? Reading and calculation have been with us only a few thousand years, much too short a lapse of time for evolution to have instilled in us a genetic predisposition for these functions. Such cognitive abilities of recent origin must therefore invade cerebral circuits initially assigned to a different use. They take them over so thoroughly that they seem to become the circuit's new dedicated function.

The basis for such changes in the function of cerebral circuits is *neuronal plasticity*: the ability of nerve cells to rewire themselves, both in the course of normal development and learning, and following brain damage. Neuronal plasticity, however, is not unlimited. In the final analysis, the adult pattern of cerebral specialization must therefore result from a combination of genetic and epigenetic constraints. Certain regions of the visual cortex, initially involved in object or face recognition, progressively become specialized for reading when a child is raised in a visual universe dominated by printed characters. Patches of cortex entirely dedicated to digits and to letters emerge, perhaps by virtue of a general learning principle ensuring that neurons coding for similar properties will tend to group together on the cortical surface. Likewise, the primate brain comprises innately specified circuits for learning and executing motor sequences. When a child acquires multiplication tables, these circuits are naturally called upon and therefore tend to specialize for calculation. Learning probably never creates radically novel cerebral circuits. But it can select, refine, and specialize preexisting circuits until their meaning and function depart considerably from those Mother Nature initially assigned them.

Flagrant limits to cerebral plasticity are seen in children who suffer from *developmental dyscalculia,* a seemingly insurmountable deficit in arithmetic acquisition. Some of these children, although their intelligence is normal and they obtain good results at school in most subjects, suffer from an exceedingly narrow handicap that recalls the neuropsychological deficits seen in brain-damaged adults. The odds are that they were subject to a precocious neuronal disorganization within cerebral areas that should have normally specialized for number processing. Here are three remarkable examples brought to us by the English neuropsychologist Christine Temple and the psychologist Brian Butterworth:

- S. W. and H. M. are teenagers of normal intelligence who attend a conventional school. Both speak fluently. H. M. is dyslexic, but her reading handicap does not extend to numbers: Like S. W., she can read Arabic numerals aloud and compare them. Yet H. M. and S. W. exhibit a double dissociation within calculation. S. W. knows his arithmetic tables to near perfection and can add, subtract, or multiply any two digits. However, he repeatedly fails in multidigit calculations: He errs in the order and nature of the component operations, and he carries over without rhyme or reason. Since childhood, he has suffered from a selective deficit of calculation procedures so severe that even a specialized rehabilitation program has not been able to compensate it. Conversely, H. M. is a master in multidigit calculation algorithms, but she could never learn the multiplication table. At nineteen, she still requires more than seven seconds to multiply two digits, and the result that she reaches is incorrect in more than half of the trials.

  S. W. and H. M.'s highly selective deficits are unlikely to be due to their laziness or to major flaws in their education. A neurological origin is more likely. Since childhood S. W. has suffered from tuberous sclerosis and epileptic fits. His CT scan shows an abnormal mass of nerve cells in the right frontal lobe, an anomaly that may well account for his insurmountable inability to perform sequential calculations. As to H. M., although she suffers from no known neurological disorder, it would be well worth examining with modern brain imaging tools the extent to which her parietal lobe and subcortical circuits are intact.

- Paul is an eleven-year-old boy of normal intelligence. He suffers no known neurological disease, has a normal command of language, and uses an extensive vocabulary. Yet from his earliest youth Paul has experienced exceptionally severe difficulties in arithmetic. Multiplication, subtraction, and division are impossible to him. At his best, he occasionally succeeds in adding two digits by counting on his fingers. His deficit even extends to reading and writing numbers. When taking numbers down in dictation, instead of 2 he writes down 3 or 8!.He also fails dramatically when reading Arabic numerals or spelled-out number words aloud: 1 is read as "nine," and *four* as "two." Only numerals are subject to these strange word substitutions. Paul can read even the most complex and irregular English words, such as *colonel*. He even finds a plausible pronunciation for fictitious words such as *fibe* or *intertergal*. Why then does he read the word *three* as "eight"? Paul apparently suffers from a complete disorganization of number sense, comparable in severity to Mr. M's predicament. This deficit occurred so early on that it seems to have prevented Paul from attributing any meaning to number words.

- C. W. is a young man in his thirties. His intelligence is normal, although he never really shone in school. Though he can more or less read and write numerals under three digits long, their quantitative meaning escapes him. Adding or subtracting two digits takes him more than three seconds. In order to multiply, he resorts to repeated addition. He succeeds only when both operands are smaller than 5 and can therefore be portrayed with the fingers of one hand. More surprising still, he cannot tell without counting which of two numbers is the larger. He thus shows an *inverse* distance effect: in contrast to a normal person, it takes him *less* time to compare 5 and 6 than to compare 5 and 9, because the larger the numerical distance, the longer he has to count. Even the subitizing of very small sets of objects is beyond his reach. When three dots appear on a computer screen, he has no immediate notion of their numerosity unless he counts them one by one. C. W. seems to have been devoid from childhood of any rapid and intuitive perception of numerical quantities.

These remarkable cases call into question the extent of cerebral plasticity in the developing brain. Although neuronal circuits are highly modifiable, especially in young children, they are not ready to assume any function. Some circuits, whose main connection patterns are under genetic control, are biased to become the neuronal substrate of narrowly defined functions such as the evaluation of numerical quantities or the storage of rote multiplication facts. Their destruction, even in the very young, can cause a selective deficit that is not always open to compensation by neighboring brain areas.

This observation brings us back once more to a recurrent theme in this book: the strong constraints that our cerebral architecture imposes on the mental manipulation of mathematical objects. Numbers do not have full latitude to invade any available neuronal networks of the child's brain. Only certain circuits are capable of contributing to calculation—either because they are part of our innate sense of numerical quantities, such as, perhaps, some areas of the inferior parietal cortex, or because, though they were initially destined for some other use, their neural organization turns out to be sufficiently flexible and close to the desired function so that they can be "recycled" for number processing.

Chapter 8

The Computing Brain

> A picture . . . shows Einstein lying in bed, his head bristling with electrical wires: his brain waves are being recorded while he is asked to "think about relativity."
>
> Roland Barthes, *Mythologies*

Nobel Prize winner Richard Feynman once remarked that the physicist who analyzes subatomic collisions in a particle accelerator is not very different from someone who sets out to study clockmaking by smashing two watches together and examining the remains. This tongue-in-cheek remark applies equally well to neuropsychology. It too is an indirect science in which the normal organization of cerebral circuits is inferred from the way they function after having been damaged—an awkward enterprise not unlike trying to deduce the inner workings of a clock from the examination of hundreds of broken movements.

Even if most brain scientists trust neuropsychological inferences, there comes a time when they would like to "open the black box" and observe the neural circuits underlying mental calculation directly. It would be an extraordinary step forward if we could somehow measure the cellular firing patterns that code for numbers. Jean-Pierre Changeux maintains this forcefully: "These 'mathematical objects' correspond to physical states of our brain in such a way that it ought *in principle* to be possible to observe them from the outside looking in, using various methods of brain imaging."

This neurobiologist's dream is now on the verge of becoming a reality. In the last few years, new tools—positron emission tomography, functional magnetic resonance imaging, and electro- and magnetoencephalography—have begun to provide pictures of brain activity in living, thinking humans. With modern brain imaging tools, a short experiment is now sufficient to examine which brain regions are active while a normal subject reads, calculates, or plays chess. Recordings of the electrical and magnetic activity of the brain with millisecond accuracy allow us to unveil the dynamics of cerebral circuits and the precise moment when they become active.

In several respects, the new pictures of the active brain are complementary to the results gathered from neuropsychology. Several cerebral areas long failed to be appreciated by neuropsychologists, either because they were rarely lesioned or because their destruction was very damaging or lethal. Today an entire network can be visualized in a single experiment. In the past, it was also difficult to study the temporal organization of cerebral circuits in the damaged brain, which often undergoes a profound reorganization. Modern imaging is able to disclose the propagation of neuronal activity to many successive regions of the normal human brain, almost in real time.

We now have amazing equipment worthy of an Isaac Asimov novel at our disposal. How can one fail to marvel at the idea that we can visualize the physiological changes that support our thoughts? Since this new world has been made accessible to scientists, dozens of experiments have explored the cerebral basis of functions as diverse as reading, motion perception, verbal associations, motor learning, visual imagery, and even our sense of pain. It would be impossible to review in full all the discoveries that this methodological revolution has permitted. In this chapter, I focus exclusively on studies that reveal human cerebral activity during mental arithmetic.

Does Mental Calculation Increase Brain Metabolism?

To retrace the heroic beginning of brain imaging, we must temporarily forget all about modern technologies and head far back into the history of neuroscience. In 1931, a report by William G. Lennox from the department of neuropathology at Harvard, soberly titled "The cerebral circulation: the effect of mental work," was the first to boldly probe the impact of arithmetic activity on brain function. Lennox raised the critical issue of the influence of cognitive processing on the energy balance of the brain. Does mental calculation involve a measurable

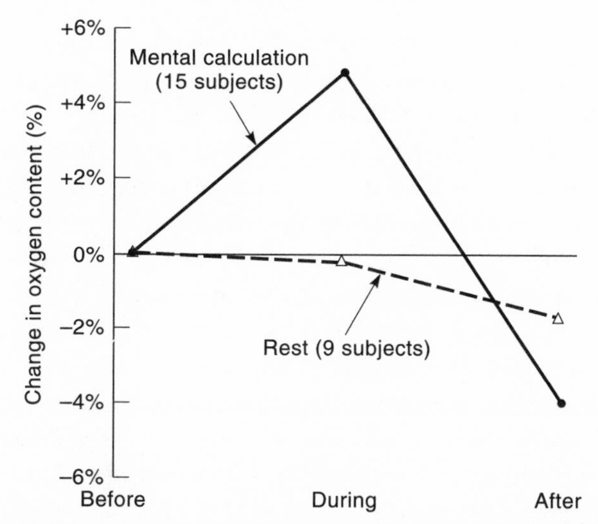

Figure 8.1. *As early as 1931, William Lennox showed that intense mental calculation changes the oxygen content of blood samples taken from the internal jugular vein. (After Lennox, 1931.)*

expenditure of energy? Does the brain burn more oxygen when the computations it performs increase in intensity?

The experimental method that Lennox devised was innovative but appalling. It consisted of drawing blood samples from the internal jugular vein and measuring their oxygen and carbon dioxide content. The article did not report whether the twenty-four subjects, epileptic patients who were being treated at Boston City Hospital, had been informed of the risk that they incurred and of the non-therapeutic objectives of the research. In the 1930s, ethical standards were still quite lenient.

Lennox's experimental design, however, was clever. In a first group of fifteen subjects, he took three consecutive blood samples. The first was taken after the subjects had rested for half an hour with their eyes closed. They were then given a sheet covered with arithmetic problems, and five minutes later, while they struggled to solve them, a second blood sample was taken. Finally, the subjects were allowed to rest for ten to fifteen minutes before the final sample was taken. The results are striking: Among the three measures, the one that had been performed during mental calculation showed a marked increase in oxygen content (Figure 8.1). Lennox did not report any statistical test on this finding, but my own calculations from the raw data evaluate to only about 2% the likelihood that this large variation across samples could be due to chance.

One objection, however, had to be refuted. That was, in the author's own words, "It is difficult for the subject either to 'make his mind a blank' or to concentrate on the problems set before him while a needle is being inserted deeply into his neck (sic!). The degree of apprehension or of discomfort may not have been the same each time that blood was withdrawn."

To meet this criticism, Lennox took the precaution of repeating the same series of three measures on another group of nine subjects that remained at rest throughout the test. For these subjects, the oxygen content remained practically constant. Thus, the intense efforts required by mental calculation had to be responsible for the increase observed in the experimental group. The finding opened revolutionary perspectives. For the first time, an objective measurement of the energy consumed by intellectual effort could be envisaged.

In detail, however, the results raised an apparent paradox that Lennox did not fail to notice. Blood was drawn from the internal jugular vein, hence *after* it had irrigated the cerebrum. But mental activity was expected to increase the consumption of oxygen. Thus, for equal cerebral blood flow, the oxygen content of the venous blood should have decreased rather than increased during intellectual work. To resolve this contradiction, Lennox exhibited remarkable anticipation powers by stating, as early as in 1931, a principle that has remained valid to this day: "The result can be explained by a dilatation of cerebral vessels with a resulting increase in the speed of blood flow through the brain, a factor that outweighs the increased consumption of oxygen."

The most recent studies in functional brain imaging have confirmed this postulate, which lies at the heart of the modern method of functional magnetic resonance imaging. The regulation system that accelerates cerebral blood flow in response to a local increase in neuronal activity does indeed bring in more oxygen than the brain can consume. The reasons for this curious phenomenon still remain poorly understood. That Lennox managed to foresee it shows the extent to which one can trust his work, despite the primitive and invasive technique on which it is based.

To close this historical discussion, it should be noted that a subsequent study by Louis Sokoloff and his colleagues at the University of Pennsylvania, in 1955, did not manage to replicate Lennox's results (though it relied on a slightly different method). Looking back on it, several criticisms also come to mind. First, the increase in the oxygen content that Lennox observed may have had little to do with mental calculation. It could simply have been due to the intense perceptual and motor activity required to scrutinize a sheet filled with mathematical signs and to give the numerical results. In other words, nothing proves that Lennox really measured the physiological bases of a purely *mental* activity as opposed to greater visual or motor work.

To a modern reader, however, the article's most obvious deficiency lies in its total neglect of the issue of cerebral localization. During calculation, does cerebral blood flow increase throughout the cerebrum? Or are the changes circumscribed to specific brain regions? And in the latter case, could cerebral blood flow serve as a tool to localize areas dedicated to distinct mental processes on the cortical surface? Lennox's article did not even mention whether the blood samples had been drawn from the left or the right internal jugular vein, a fact that might have supported conclusions about the hemispheric lateralization of mental calculation procedures. Improvements in spatial localization, and the production of genuine pictures of human brain activity, were obliged to wait until the 1970s and 1980s, which finally saw the advent of reliable functional brain imaging techniques.

The Principle of Positron Emission Tomography

Following Lennox's pioneer work, several studies have confirmed that the brain is amazingly voracious in its need for energy. Indeed, it alone is responsible for almost a quarter of the energy expended by the entire body. Its local energy consumption, however, is not a constant. It can suddenly rise, in a matter of seconds, when a cerebral region is put to use. Sokoloff was the first to demonstrate the direct relations between cerebral blood flow, local metabolism, and the degree of activity of cerebral areas. If I decide to rapidly wiggle my right index finger, for instance, neurons start to fire within the minuscule patch of left motor cortex dedicated to the command of the muscles driving that finger. A few seconds later, glucose consumption increases in this area of cerebral tissue. In parallel, cerebral blood flow increases within the vessels and capillaries that irrigate the region. The increased volume of circulating blood meets and even exceeds the local increase in oxygen consumption.

In the last twenty years, these regulation mechanisms have been exploited to determine which brain regions are active during various mental activities. At the heart of these innovative brain imaging techniques is an extremely simple idea: If one can measure local glucose metabolism or blood flow in a given brain area, one should immediately obtain an indication of recent neural activity. But the implementation of this idea is tricky. How can one assess blood flow or the quantity of degraded glucose at each point in the brain?

Sokoloff found a solution for animals. His now classical autoradiography technique consists in injecting a molecule marked with a radioactive tracer, such as fluorodeoxyglucose, and then having the animal perform the desired task (say, moving the right paw). The radioactive fluorine atom, attached to the glucose molecule, is preferentially deposited in the cerebral regions that burn up the most energy. Subsequently, the animal's brain is cut into thin slices. Each slice is placed in darkness against a photographic film, which gets exposed only directly opposite the zones where radioactivity is concentrated. The series of slices thus permits reconstruction of the three-dimensional extent of the areas that were active at the time of the injection.

The spatial resolution of autoradiography is excellent, but obvious reasons make it unsuitable for research with humans: Neither brain slicing nor the injection of high doses of radioactivity are likely to meet the subject's approval. These difficulties can be circumvented, however, by the magic of three-dimensional reconstruction methods derived from physics and computer science. Experiments with humans only use radioactive tracers with a short half-life span, anywhere between a few minutes and a few hours. As soon as the experiment is over, all radioactivity quickly vanishes. The injected doses of radioactivity are harmless unless exposure is repeated frequently. Thus the experiment is no more dangerous for the subject than the typical X-ray, and no more painful than a regular intravenous injection. In order for the experiment to proceed in accordance with medical ethics, subjects are fully informed of the objectives and methods used in the research before they volunteer.

Only one problem remains: how to detect the concentration of radioactivity within the physically inaccessible volume of the skull? Positron emission tomography, also known as "PET scanning," provides a high-tech solution. Consider for a moment the nuclear physics of a subject who has just been injected with a tracer that emits positrons—for instance, a water molecule in which the usual oxygen atom has been replaced by an unstable atom of oxygen 15 ($H_2{}^{15}0$). After an unpredictable delay ranging from seconds to minutes, this atom emits a positron, an antimatter particle denoted as e^+ whose properties are exactly symmetrical with those of the familiar electron e^-. The subject's head, his whole body in fact, is thus turned into an antimatter generator! As you may guess, this state of affairs cannot last very long. Only a few millimeters away, the positron collides with its twin the electron, which abounds in normal matter. The two annihilate each other by emitting two high-energy gamma rays of opposite polarization that exit from the scalp without interacting much with the surrounding atoms.

The secret of PET scanning consists in detecting the photons emitted by the subject's brain. To this end, hundreds of crystals coupled to photomultipliers are arranged into a circle around the head, and they detect any suspect disintegration. In the older technique of single photon emission tomography, only isolated gamma rays emitted by a radioactive source such as Xenon (^{133}Xe) were of interest. In positron emission tomography, it is the simultaneous occurrence of two gamma rays that is sought. The quasi-simultaneous detection of two photons by diametrically opposite detectors is an almost sure sign that a positron has disintegrated. The alignment of the detectors, sometimes combined with an analysis of the minuscule lag between the two detections ("time of flight"), helps locate this disintegration in all three dimensions. As indicated by its etymology, the tomograph thus produces a "sliced picture" of the distribution of radioactivity in a given volume of brain tissue. This quantity of radioactivity is a good indicator of local cerebral blood flow, which is itself a good indicator of the average neuronal activity in that area.

Practically speaking, a typical experiment using positron emission tomography runs as follows: A volunteer, lying in the tomograph, starts to perform the requested task (moving the index finger, multiplying digits, etc.). At the same time, a cyclotron produces a small quantity of a radioactive tracer. As soon as it is available, the tracer must be injected immediately or else its radioactivity rapidly decreases below the detectable level. The subject continues mental activity for one or two minutes after the injection. Throughout that period, the tomograph reconstructs the spatial distribution of radioactivity in the subject's brain. The volunteer then rests for ten to fifteen minutes until radioactivity falls back to an undetectable level. The procedure can then be repeated up to twelve times in the same subject, possibly with different task instructions on each injection.

Can One Localize Mathematical Thought?

Although the first pictures of the active brain date back to the 1970s, our quest for images of the calculating brain takes us back only as far as 1985. That year, two Swedish researchers, P. Roland and L. Friberg, published a result that fills many of the gaps left by Lennox's work. The first sentences of their article sets the framework:

> These experiments were undertaken to demonstrate that pure mental activity, thinking, increases the cerebral blood flow and that different types of thinking increase the regional cerebral blood flow in different cortical areas. As a first approach, thinking was defined as *brain work in the form of operations on internal information, done by an awake subject.*

In order to pinpoint "thought processes," Roland and Friberg meticulously controlled the tasks that they asked subjects to perform. In the task most relevant to this discussion, the subjects had repeatedly to subtract 3 from a given number (50–3=47, 47–3=44; etc.). The calculations were silent. Only after a few minutes did the experimenter interrupt the subjects and asked them to say what number they had reached. Throughout the measurement interval, mental operations thus pro-

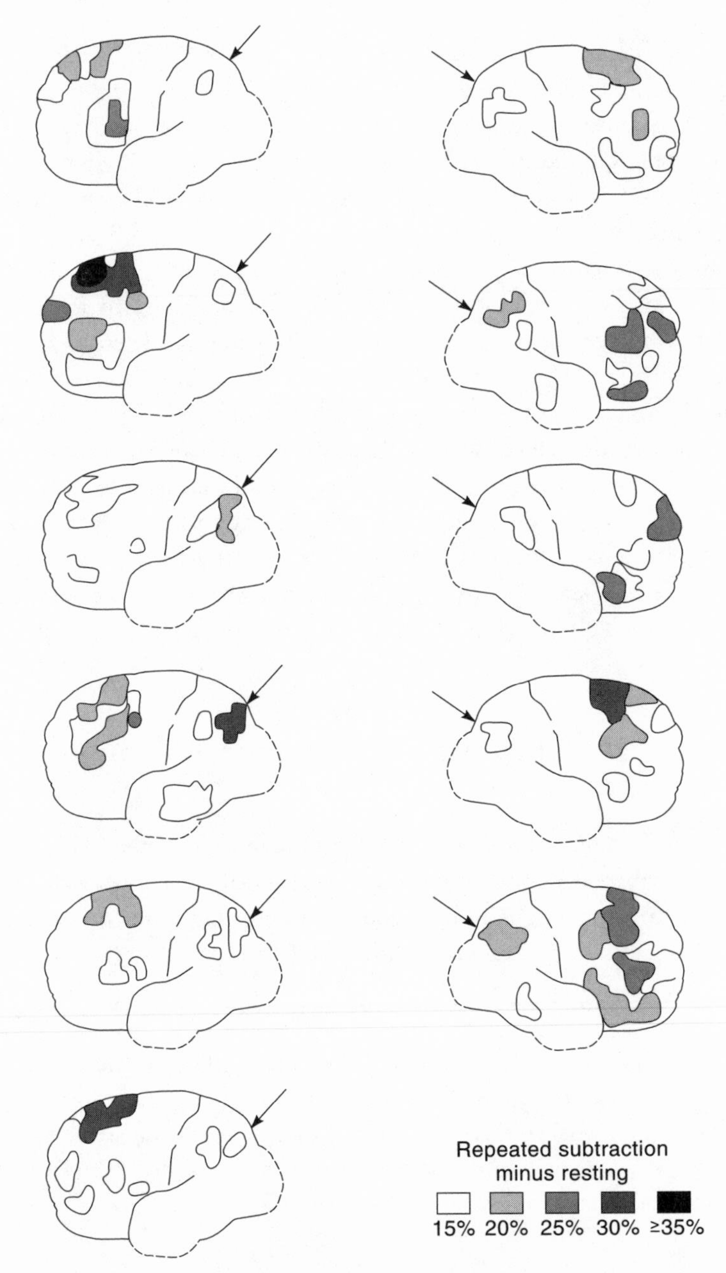

Figure 8.2. *In 1985, Roland and Friberg published the first images of cerebral activity during mental calculation. At the time, their method could only visualize one hemisphere at a time. Each image thus represents the data from one volunteer. When compared to a rest period, repeated subtraction yields bilateral activations in the inferior parietal cortex (arrow) as well as in multiple regions of the prefrontal cortex. (Adapted from Roland and Friberg 1985; copyright © 1985 by American Physiological Society).*

ceeded in a purely internal manner, with no detectable sensory or motor activity.

In addition to this mental calculation task, two other tests studied either spatial imagery (picture in your mind the route that you would follow if you left home and took alternate right or left turns) or verbal flexibility (mentally recite a word list in an unusual order). The brain regions that were active during each task were determined by comparison to a measure of cerebral blood flow obtained while the subject was at rest, thinking about nothing in particular. The brain imaging procedure used by Roland and Friberg, now outdated, called for an injection of radioactive Xenon (^{133}Xe) in the internal carotid artery and the detection of single photons.Without attaining the accuracy of PET scanning, the method visualized local increases in blood flow near the cortical surface.

In each of eleven volunteers, the cerebral activations during mental calculation were concentrated in two major brain areas: a vast prefrontal region and a most restricted inferior parietal region near the angular gyrus (Figure 8.2). Both regions were found active in the left and right hemispheres, although the activation was slightly greater on the left than on the right.

The anatomical precision of this early experiment was far from perfect, but its results have recently been replicated by Jordan Grafman, Denis Le Bihan, and their colleagues at the National Institutes of Health with a much more accurate method called functional magnetic resonance imaging. Bilateral activations of the prefrontal and inferior parietal cortices were again found in all subjects during repeated subtraction, although the number of activated pixels was larger in the left hemisphere than in the right. I served as a subject in a very similar pilot experiment in Orsay, near Paris. Figure 8.3 shows a slice of my brain while I struggle with the repeated subtraction task. Bilateral parietal and prefrontal activations are clearly visible.

The results from the other conditions of Roland and Friberg's experiment suggested that parietal and prefrontal activations were related to different aspects of the task. The prefrontal region was found in all mental manipulation tasks, not just those involving mental subtraction. Roland and Friberg ascribed it a very general role in the "organization of thought." By contrast, the inferior parietal region seemed specific to mental calculation, since it did not activate during spatial imagery or verbal flexibility tasks. The two researchers attributed to it a specialization for mathematical thought, and in particular for the retrieval of subtraction results from memory.

Roland and Friberg's experiment played a crucial role in drawing the attention of the scientific community to the power of functional imaging, a full three years before Michael Posner, Steve Petersen, Peter Fox, and Marcus Raichle's celebrated demonstration of distinct brain activations for different aspects of language processing. The Swedish team's work indeed proved that the new techniques

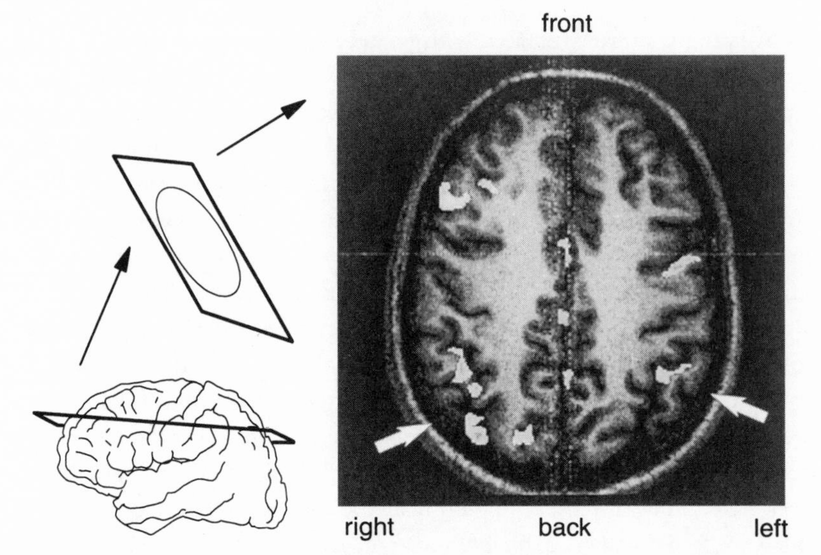

Figure 8.3. A slice through the author's brain during a replication of Roland and Friberg's experiment. Cerebral regions whose activity increases whenever I subtract were determined by high-field (3 Tesla) functional magnetic resonance imaging and were superimposed on a classical anatomical MR image. Activations are visible in the inferior parietal cortex (white arrows) and the prefrontal cortex. (Dehaene, Le Bihan, and van de Moortele, unpublished data, 1996.)

could resolve brain activation differences related to distinct cognitive tasks. What should one make, however, of their general conclusions concerning "thinking"? Can one really localize a cerebral area of "mathematical thought" in the human brain?

Personally, I take Roland and Friberg's functional labels with a pinch of salt. The very notion that "thought" is a valid object of scientific study and that it can be localized to a small number of cerebral areas recalls an old discipline that was once relegated to the museum but is making an insidious comeback: Gall and Spurzheim's phrenology, or the hypothesis that the brain contains a panoply of organs, each dedicated to a very complex function such as the "love of one's progeny." Phrenology has been abandoned for more than a century. It would be surely unfair to accuse Roland and his colleagues, who pioneered the field of brain imaging, of trying to revive it. Yet it takes little sagacity to observe that many recent experiments in brain imaging are conceived in a "neo-phrenologic" framework. Their only objective seems to be the labeling of cerebral areas. Positron emission tomography is implicitly treated by many research groups as a simple mapping tool that directly discloses the cerebral areas underlying a given function, be it mathematics, "thought," or even consciousness. The method supposes a clear and unique relationship between cerebral areas and cognitive abilities: Calculation rests on the inferior parietal region, the organization of thought is taken care of by frontal cortex, and so on.

We have every reason to think that the brain does not work this way. Even seemingly simple functions call for the coordination of a large number of cerebral areas, each making a modest and mechanical contribution to cognitive processing. Ten or twenty cerebral areas are activated when a subject reads words, ponders over their meaning, imagines a scene, or performs a calculation. Each region is responsible for an elementary operation such as recognizing printed letters, computing their pronunciation, or determining the grammatical category of a word. Neither an isolated neuron, nor a cortical column, nor even a cerebral area can "think." Only by combining the capacities of several million neurons, spread out in distributed cortical and subcortical networks, does the brain attain its impressive computational power. The very notion that a single cerebral region could be associated with a process as general as the "organization of thought" is now obsolete.

How, then, should one reinterpret Roland and Friberg's results? As we saw in Chapter 7, the inferior parietal area is the region that is impaired in Gerstmann's syndrome. Damage to it was responsible for the loss of number sense in patient Mr. M, who was so impaired that he could no longer compute 3–1 and believed that 7 fell between 2 and 4. Hence, this region probably contributes to a narrow process: the transformation of numerical symbols into quantities, and the representation of relative number magnitudes. It does not play a generic role in arithmetic since damage to it does not necessarily affect the rote retrieval of simple arithmetic facts (2+2=4), nor the rules of algebra ($(a+b)^2 = a^2 + 2ab + b^2$), nor the encyclopedic knowledge of numbers (1492=Columbus's discovery). It is involved only in the representation of numerical quantities and their positioning on a mental number line. Its activation during repeated subtraction in normal subjects thus provides a nice confirmation of its crucial role in processing quantities.

As for the extended prefrontal activation reported by the Swedish team, it probably embraces several areas, each with its own function: sequential ordering of successive operations, control over their execution, error correction, inhibition of verbal responses, and, above all, working memory. In a sector of prefrontal cortex called the dorso-lateral region or "area 46," neurons are known to be involved in the on-line maintenance of past or anticipated events in the absence of any external input (as when we rehearse a phone number, for instance). Remarkable experiments by Joachim Fuster and Patricia Goldman-Rakic, among others, have shown that prefrontal cortical neurons maintain a sustained level of firing when a monkey holds information in memory for several seconds. All three tasks employed by Roland and Friberg relied heavily on this type of working memory. In the repeated subtraction task, for instance, subjects had constantly to keep in mind the number that they had reached and update it after each subtraction. This important memory load likely explains the involvement of prefrontal circuits in this task.

When the Brain Multiplies or Compares

Roland and Friberg's experiment probed only a single complex arithmetical task, with the aim of identifying the areas involved in arithmetic. This was just a first step. Neuropsychological dissociations lead us to expect a much finer-grained fragmentation of cerebral areas. Depending on the requested arithmetic operation, very different cerebral networks should activate. To begin to evaluate this hypothesis, my colleagues and I recently examined how cerebral activity changes in the course of number comparison and multiplication.

The experiment was performed in Orsay at a medical research center well equipped for measuring cerebral metabolism. Eight medical students served as volunteers. Upon their arrival at the hospital in the morning, high-resolution magnetic resonance anatomical images of their brains were made. Later in the afternoon, positron emission tomography provided us with the first detailed images of the areas that were activated while they processed numbers.

Remember Mr. N, the patient who could not multiply but could still tell which of two numbers was larger? The goal of our study was to investigate whether the neuronal circuits involved in multiplication and comparison did partially rest on distinct brain areas, as we had postulated based on Mr. N's results. We thus presented subjects with a series of pairs of digits that they either had to compare or multiply mentally. In both cases, the result of the operation—either the larger of the two digits or their product—had to be named covertly, without actually moving the lips. Cerebral blood flow during those two tasks was contrasted to a third measure obtained while the subjects were at rest.

As we expected, several brain regions were equally active during multiplication and during comparison relative to the rest period. These regions most probably support functions common to both tasks, such as extracting visual information (occipital cortex), or maintaining gaze fixation and the internal simulation of speech production (supplementary motor area and precentral cortex).

The inferior parietal cortex, so crucial to quantitative number sense, was also active. Oddly, it was intensely active in both hemispheres during multiplication, while its activity during comparison was small and on the verge of being inde-

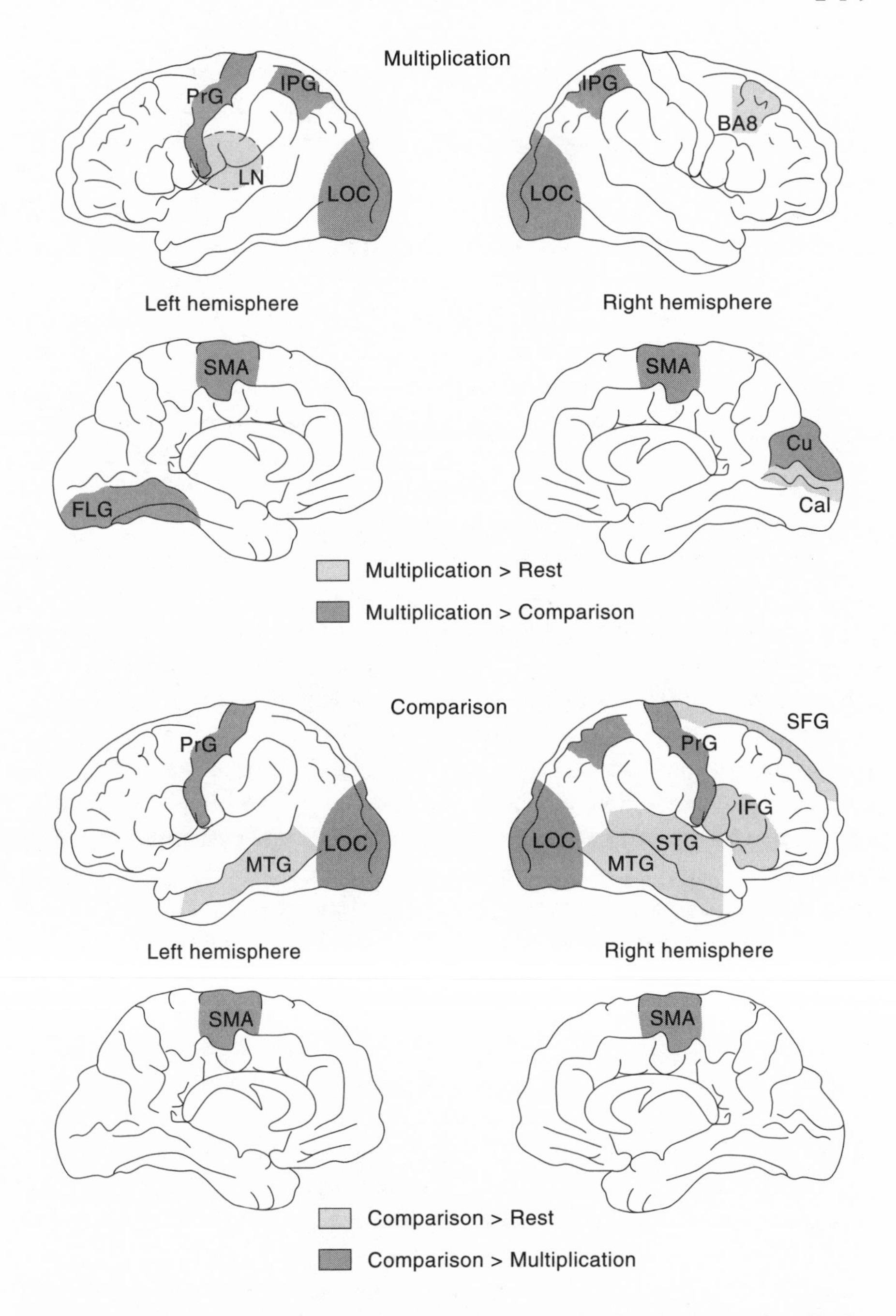

Figure 8.4. Positron emission tomography reveals wide networks of cerebral areas whose blood flow changes when subjects rest with their eyes closed, multiply pairs of Arabic digits, or compare the very same digits. (After Dehaene et al. 1996.)

tectable. We had expected the reverse: Comparison calls for the processing of numerical quantities, and simple multiplication requires only access to verbal memory. However, not all the multiplication problems we used were simple. The list included problems such as 8×9 or 7×6 on which our subjects often hesitated or failed altogether. Since their verbal memory for arithmetic facts seemed unreliable, we speculate that they were often forced to resort to backup strategies heavily dependent on the inferior parietal cortex to provide a plausible answer. Conversely, the number comparison task we used was probably too easy becausee the numbers ranged only from 1 to 9. Finding the larger digit may have been too simple to stimulate intense inferior parietal activation. Perhaps we also left the subjects too much time to respond, which may have diluted the activations to the point of rendering them too small to detect. At any rate, inferior parietal cortex seemed to activate in direct proportion to the difficulty of the numerical tasks that the subjects performed.

The most interesting results emerged, however, when we directly contrasted number comparison with multiplication. Several temporal, frontal and parietal regions showed a notable shift in hemispheric asymmetries. During multiplication, cerebral activity was more intense in the left hemisphere, but during comparison it was equally distributed across the two hemispheres or even shifted to the right. This observation is in agreement with the notion that multiplication, but not comparison, rests in part on the language abilities of the left hemisphere. Contrary to multiplication, number comparison does not have to be learned by rote. A mental representation of number magnitude emerges without explicit teaching in young children and even in animals. Hence, the brain does not need to convert digits to a verbal format in order to compare them. Functional brain imaging confirms that the comparison of numerical magnitudes is a nonlinguistic activity that rests at least as much on the right hemisphere as on the left. Each hemisphere can recognize digits and translate them into a mental representation of quantities to compare them.

A subcortical nucleus, the left lenticular nucleus, was also more active during multiplication than during comparison. We know from Chapter 7 that a lesion in this area can dramatically impair memory for multiplication facts and other verbal automatisms. Remember Mrs. B, who had forgotten how to recite "three times nine is twenty-seven," the alphabet, and the Our Father? Her lesion was right in this area. The lenticular nucleus belongs to the basal ganglia, which are generally thought to contribute to the routine aspects of motor behavior. Functional brain imaging suggests that they also contribute to more elaborate cognitive functions. Perhaps arithmetic tables are stored in the form of automatic word sequences, so that recalling them becomes mechanical. Reciting the multiplication table at school may imprint every word of it in our deep brain structures. This would explain why even the most fluent bilinguals still prefer to calculate in the language in which they acquired arithmetic.

The diversity of cerebral areas involved in multiplication and comparison underlines once more that arithmetic is not a holistic phrenological "faculty" associated with a single calculation center. Each operation recruits an extended cerebral network. Unlike a computer, the brain does not have a specialized arithmetic processor. A more appropriate metaphor is that of a heterogeneous group of dumb agents. Each is unable to accomplish much alone, but as a group they manage to solve a problem by dividing it among themselves. Even an act as simple as multiplying two digits requires the collaboration of millions of neurons distributed in many brain areas.

The Limits of Positron Emission Tomography

Positron emission tomography is a wonderful tool, but it has some unfortunate limits. To verify our hypotheses on the cortical and subcortical processing of numerical information, we would ideally like to observe the time course of cerebral activations during calculation. If possible, we would want to obtain a new image of brain activity every hundredth of a second. We could then follow the propagation of neuronal activity from the posterior visual areas all the way to the language areas, the circuits controlling memory, the motor regions, and so on. Yet though PET scanning is a remarkable tool for identifying active anatomical regions, its excellent spatial resolution is accompanied by a deplorable temporal resolution. Each image depicts the average blood flow over a period of at least forty seconds. Thus, PET is almost totally blind to the temporal dimension of brain activity.

There are two main reasons for this technical limitation. First, the photomultipliers that tally up positron disintegrations must detect a minimum number of events before a significant picture emerges. Yet the number of disintegrations per second is a direct function of the dose of injected radioactivity, which, for ethical reasons, cannot be raised much beyond today's limits. Second, even if the duration of each measurement could be shortened, temporal accuracy would remain fundamentally limited by the delayed response of cerebral blood flow to a change in neural activity. When neurons in a given area start to fire, several seconds elapse before blood flow starts to rise. Even the recent technique of functional magnetic resonance imaging, which can acquire images of blood flow in a fraction of a second, suffers to a similar extent from the slowness of blood flow responses.

In a nutshell, here is the crux of the problem. The brain detects, computes, reflects, and reacts in a fraction of a second. Functional techniques based on blood flow reduce this complex sequence of activity to a static picture. It is comparable to photographing the finish of a horse race with an exposure time of several seconds. The fuzzy picture might show which horses made it past the finish line, but the order in which they arrived would be lost. What we need is a technique that could take a series of snapshots of cerebral activity and later replay the movie in slow motion.

The Brain Electric

Electro- and magnetoencephalography are the only techniques that currently come close to meeting this challenge. Both take advantage of the fact that the brain behaves like a generator of electric current. To better understand how they work, a quick reminder of how nerve cells communicate might help. Any nervous system, whether it belongs to a human or a leech, consists primarily of a packed bundle of cables. Each neuron has an axon, a long cable that conveys information through waves of depolarization called action potentials. Each neuron also possesses a bushy arborization of dendrites that receive the signals coming from other nerve cells. When an action potential reaches a synapse—the contact zone between one neuron's axon terminal and another's dendrite—neurotransmitter molecules are released from the nerve terminal and tie on to other specialized molecules called receptors inserted within the dendritic membrane. This causes the receptors to alter their shape. They switch to an "open" configuration in which a channel opens through the cell membrane, letting ions flow into the cell. Very schematically, this is how a nerve impulse crosses the barrier of the cellular membrane and is transmitted from one neuron to the next.

Since ions carry an electric charge, their movement across the cellular membrane and within the dendritic tree produces a very small amount of current. Each neuron thus behaves as a tiny electric generator. Indeed, the electric organ of fish such as the torpedo ray is nothing but a giant synapse in which such electrochemical units are arranged into a powerful battery. From the torpedo's electrical organ to the human nervous system, the molecular mechanism is so similar that an almost identical receptor molecule is found in both. Molecular neurobiologists were thus able to make an important step forward when a concentrate of torpedo fish provided a sufficient amount of the receptor to characterize its molecular structure.

Coming back to the human brain, each active cerebral area thus produces an electromagnetic wave form that is transmitted by volume conduction all the way to the scalp. More than fifty years ago, Hans Berger first put this knowledge into practice by affixing electrodes on the scalp of several volunteers and recording an electric signal—the first electroencephalogram. This signal, which results from the synchronous activation of several million synapses, is very weak: only a few millionths of a volt. It is also highly chaotic and shows seemingly random oscillations. However, when one synchronizes the recording with an external event, such as a visually presented digit, and when one averages across many presentations, a reproducible sequence of electric activity called the *event-related potential* emerges from the chaos. This sequence conceals a wealth of temporal information. The signals are propagated almost instantaneously to the scalp surface, where they can be recorded in real time—for instance, every millisecond. A continuous record of cerebral activity is then available, which faithfully reflects the order in which each brain region was activated.

Modern technologies now make it possible to record event-related potentials from up to 64, 128, or even 256 scalp electrodes. Their shape varies from electrode to electrode, and this spatial distribution provides precious indications about the location of active brain areas. In this respect, however, the method remains unsatisfactory. The anatomical accuracy of electroencephalographic recordings is poor because a fundamental physical ambiguity precludes their direct attribution to an identifiable anatomical structure. At best, the approximate state of activity of an extended cortical region can be reconstructed by making more or less plausible inferences. A similar difficulty affects the slightly more precise but considerably more expensive method of magnetoencephalography, in which one records magnetic fields rather than electric potentials. Both methods, however, possess an unsurpassed capacity to determine the exact time when different cerebral areas come into play during mental computations.

The Time Line of the Number Line

Any of us takes about four-tenths of a second to decide whether a given digit is larger or smaller than 5. Yet this time corresponds to the total duration of a whole series of operations, from the visual identification of the target digit to the motor response. Can it be decomposed into small steps? Electroencephalography turns out to be an ideal method for measuring, with millisecond accuracy, how long it takes our brain to decide that 4 is smaller than 5.

In one of my recent experiments, Arabic digits or number words were flashed on a computer screen. Volunteers were asked to press one key for numbers small-

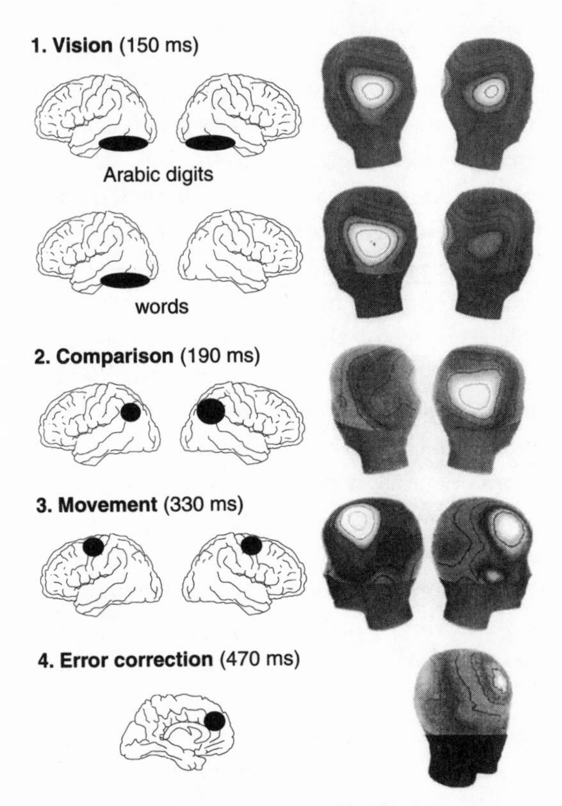

Figure 8.5. By recording the minute changes in scalp voltage generated by cerebral activity (electroencephalography), the sequence of cerebral activations during numerical comparison can be reconstructed. In this experiment, volunteers pressed keys with their left or right hand, as fast as they could, to indicate whether the numbers they saw were larger or smaller than 5. At least four processing stages were identified: 1. visual identification of the target Arabic digit or number word; 2. representation of the corresponding quantity and comparison with the memorized reference; 3. programming and execution of the manual response; and 4. correction of occasional errors. (After Dehaene 1996.)

er than 5, and another for numbers larger than 5. Their event-related potentials were recorded from sixty-four electrodes spread out on the scalp. Special software allowed for the reconstruction, frame by frame, of the evolution of surface potentials in the various conditions of the experiment (Figure 8.5).

The movie starts at the exact moment when the number appears before the subject's eyes. For several tens of milliseconds, the electric potentials remain close to zero. At around 100 milliseconds, a positive potential called the P1 appears on the rear of the scalp. It reflects the activation of visual areas of the occipital lobe. At this stage, no difference between Arabic digits and number words is perceptible: only low-level visual procedures are engaged. But suddenly, between 100 and 150 milliseconds, the two conditions diverge. While words such as *four* generate a negative potential almost completely lateralized to the left hemisphere, digits such as 4 produce a bilateral potential. As we had inferred from the performance of split-brain patients, the two hemispheres are simultaneously implied in the visual identification of Arabic digits. Number words, however, are recognized only by the left hemisphere.

Over the left-hand side at the back of the scalp, the event-related potentials evoked by words and digits appear virtually identical. More precise recordings suggest, however, that they may originate from distinct but contiguous brain regions of the left hemisphere. In some epileptic patients, neurosurgeons insert a panoply of electrodes right on the cortical surface in order to improve the spatial localization of the recordings by avoiding the deformation of electric responses by the skull. Truett Allison, Gregory McCarthy, and their colleagues at Yale University have exploited this situation to accurately record the responses of ventral occipito-temporal areas to different categories of visual stimuli such as words, digits, pictures of objects, and pictures of faces. Their results demonstrate

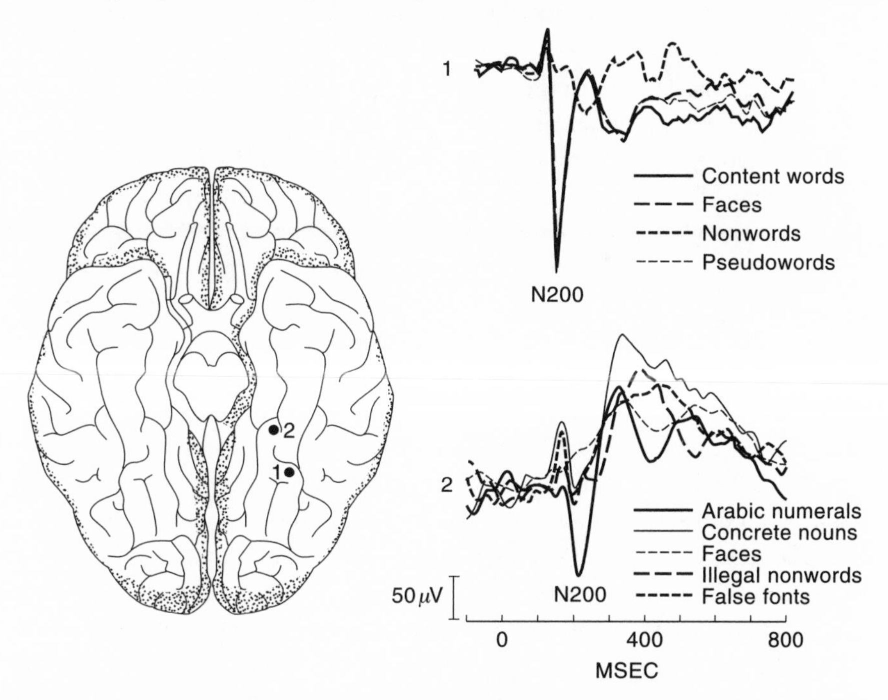

Figure 8.6. Intracranial electrodes reveal a very fine specialization of the ventral occipito-temporal region for the visual recognition of different categories of stimuli. The cortex underlying site 1 responds to letter strings (whether they spell words or not), but not to faces. A neighboring electrode at site 2 deviates only during the presentation of Arabic digits, but not of faces or letter strings. (Redrawn from Allison et al. 1994; copyright © 1994 by Oxford University Press.)

an extreme specialization. Occasionally, an electrode shows an electric deviation to words exclusively, while a second electrode one centimeter away reacts only to Arabic digits, and a third only to faces (Figure 8.6). These highly specific responses, which appear in less than 200 milliseconds, confirm that a whole collection of visual detectors, grouped according to their preferred stimuli, covers the bottom surface of the visual cortex.

Around 150 milliseconds, then, a mosaic of specialized visual areas recognizes the shape of numerical symbols. At that point, however, the brain has not yet recovered their meaning. It is only around 190 milliseconds that one sees a first indication that numerical quantity is being encoded. The distance effect suddenly emerges on electrodes located over the inferior parietal cortex. Digits that are close to 5, and therefore more difficult to compare, generate an electric potential of a greater amplitude than digits that are far from 5. The effect is seen over both hemispheres, although it is stronger on the right-hand side. It thus takes only 190 milliseconds for the brain to activate the "networks of number sense" that rest on the inferior parietal sectors of both hemispheres. Detailed analyses show that the electrical distance effect has a similar topography for Arabic digits and for number words. This confirms that the inferior parietal region is not concerned with the notation in which numbers are presented, but rather with their abstract magnitude.

Further along in our computer animation, we reach the time at which the programming of the motor response begins. An important voltage difference emerges on the electrodes located over the premotor and motor areas of both hemispheres. When subjects prepare to respond with the right hand, a negative potential appears over left-hemisphere electrodes; conversely, when they get ready for a left-hand response, it is the right-hand side of the scalp that turns negative (remember that the left motor cortex controls the movements of the right half of the body, and vice versa). This lateralized readiness potential first appears as early as 250 milliseconds after the digit first appears on screen, and it reaches its maximum around 330 milliseconds. By that time, number comparison must have been completed because the *larger* or *smaller* answer is already available. It thus takes between a quarter and a third of a second to recognize the visual shape of a digit and access its quantitative meaning.

On average, the subject's response occurs around 400 milliseconds, after an additional time lag during which the muscles contract and the subject actually executes the selected response. Yet nothing precludes continuing the analysis beyond this point. In fact, a very interesting electric event occurs right after the motor response. Even in a task as elementary as digit comparison, we occasionally make mistakes. Most errors are due to an incorrect anticipation of the response and are immediately detected and corrected. Event-related potentials betray the origin of this correction. Immediately following an error, a negative electric signal of great intensity suddenly pops up over the electrodes at the front of the skull. No such signal is found following a correct response. Hence, this activity must reflect the detection or attempted correction of the error. Its topography suggests a generator located within the anterior cingulate cortex, a cerebral area involved in the attentional control of actions and in the inhibition of unwanted behavior. Its response is so fast—less than 70 milliseconds after pressing the wrong key—that it cannot be due to feedback from sense organs. Furthermore, in my experiment no feedback was provided as to whether a response was or was not correct. The anterior cingulate cortex is thus activated in an endogenous manner whenever subjects detect that the action they are currently performing does not match the response they intend to give.

Let me stress again that all the events I have just described—number identification, access to magnitude information, comparison, response selection, execution of the motor gesture, and detection of potential errors—occur within half a second. Information passes from one cerebral area to the next with remarkable speed. At present, only electro- and magnetoencephalography provide an opportunity of following this exchange in real time.

Understanding the Word "Eighteen"

Let us consider another example of the speed of numerical information processing in the human brain. Take a look at the words EIGHTEEN and EINSTEIN. A fraction of a second suffices to notice that the first is a numeral and the second a famous physicist. It is equally easy to notice that EXECUTE is a verb, ELEPHANT an animal, and EKLPSGQI a meaningless string of letters. What cerebral areas are involved in the categorization of words of an arbitrary appearance, but with radically different meanings? Could the recording of event-related potentials reveal the activation of areas implicated in the representation of word meaning? And would the inferior parietal cortex be activated during the mere reading of the word *eighteen*, even if no calculation is required?

When volunteers pay attention to the semantic category to which words belong, scalp-recorded potentials show a remarkable sequence of cerebral activation. Initially, visual areas of the left hemisphere are equally activated by the printed strings EIGHTEEN, EINSTEIN, or EKLPSGQI. After about a quarter of a second, however, posterior visual areas discriminate actual words from meaningless strings of letters that do not obey the normal rules of word formation in English. Slightly later, around 300 milliseconds after the word appeared on screen, different categories of words also begin to diverge. Once more, numerals such as EIGHTEEN produce an electrical wave form localized in the left and right inferior parietal cortex—as if the brain had to recreate a quantitative representation of their location on the number line in order to check that these are indeed numbers.

Other word categories, by contrast, activate very different cerebral regions. Verbs, animals, and famous people all cause an extended activation of the left temporal region, which has been long-suspected of playing a special role in the representation of word meaning. Yet subtle variations appear across categories. Most notably, the names of famous persons—whether EINSTEIN, CLINTON, or BACH—are the only stimuli to activate the inferior temporal region, which other experiments have targeted in the recognition of familiar faces. Several other recent experiments suggest that this is not an isolated finding. Many categories of words—animals, tools, verbs, color words, body parts, numerals, and so on—have been found to rely on distinct sets of regions spread throughout the cortex. In each case, to determine the category to which a word belongs, the brain seems to activate in a top-down manner the cerebral areas that hold nonverbal information about the meaning of that word.

Numerate Neurons

In spite of its major contributions, electroencephalography remains an indirect and imprecise method. Tens of thousands of neurons must be activated synchronously before their electrical effect becomes detectable on the scalp. Thus, neuroscientists continue to dream of a technique that would let them examine the temporal pattern of activity of a single neuron in the human brain, as is routinely done with animals. To some extent, however, this technique is already available. Occasionally, electrodes are implanted directly into the human cortex—but the technique is so invasive that it is justified only under very exceptional circumstances. In some patients suffering from intractable epilepsy, neurosurgery is needed to remove the abnormal brain tissue from which the seizures originate. Implanting intracranial electrodes is still the best way of pinpointing the exact location of that tissue. The method consists of inserting thin needles, each with multiple electrical recording sites, deep within the cortex and subcortical nuclei. These electrodes are often left in place for several days in order to gather sufficient data about the recurring epileptic fits. With the patient's consent, nothing precludes taking advantage of this setting to study neural information processing in the human brain. Through the implanted electrodes, one can directly record electric activity in the brain while the patient reads words or performs simple calcula-

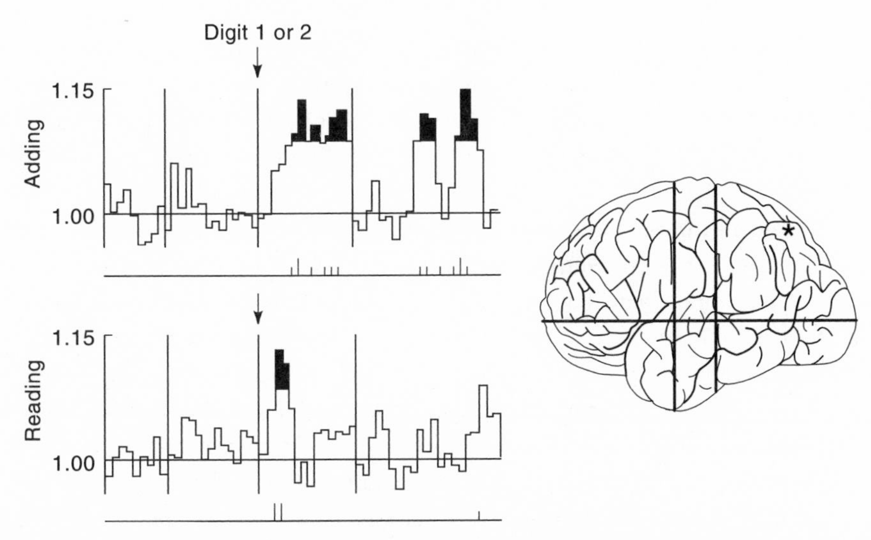

Figure 8.7. A neuron from the human parietal cortex responds selectively during number processing. The arrow indicates the time of presentation of Arabic digit 1 or 2. The intervals during which the firing frequency deviates significantly from baseline are shown in black. Neuronal activity lasts longer when the subject adds the digit to a running total than when he merely reads it aloud. (After Abdullaev and Melnichuk 1996. Courtesy Y. Abdullaev.)

tions. Depending on the characteristics of the electrode, one measures the average activity of only a few cubic millimeters of cortex, or even of a single neuron.

At the brain research center in Saint Petersburg, Yalchin Abdullaev and Konstantin Melnichuk thus recorded the activity of several single neurons in the human parietal cortex of a patient performing arithmetic and linguistics tasks. In one condition, a series of digits appeared on a screen, and the patient had to compute their running total; this was contrasted to a control situation in which the patient merely had to read the same digits aloud. In a second condition, numbers such as 54 and 7 had to be added or subtracted; again, the control consisted of reading one of the two numbers aloud. Finally, the third task, which had nothing to do with arithmetic, consisted in deciding whether a letter string such as *house* or *torse* is a valid English word or not.

The results were clear-cut. In both hemispheres, inferior parietal neurons fired only when numbers were presented. Most neurons also discharged more during calculation than during the mere reading of numbers. However, the right parietal cortex contained a few neurons whose firing frequency increased even during the reading of digits 1 and 2. When the subject was reading, these neurons fired for only a brief interval after the onset of the digit, from 300 to 500 milliseconds. But when the subject was adding or subtracting, activity lasted up to 800 milliseconds after the visual presentation (Figure 8.7).

Cellular recording thus provides direct support for the inferences we've drawn from the methods of neuropsychology, positron emission tomography, and electroencephalography. As soon as we have to manipulate numerical quantities mentally, the neural circuits of the inferior parietal cortex play an essential and very specific role.

Of course, the scattered experiments that have been reported in this chapter represent the very beginning of brain imaging. Tools for visualizing the active human brain have been around only for the last few years. Even within the domain of arithmetic, dozens of issues remained unexplored. Do parietal neurons respond specifically to certain numbers? Is the inferior parietal region organized topographically, with increasingly large numerical magnitudes systematically mapping to distinct patches of cortex? Do addition, subtraction, and comparison recruit distinct circuits? Does their organization vary with age, education in mathematics, or talent for mental calculation? To which other regions does the inferior parietal area project, and how does it communicate with the areas involved in identifying and naming words and Arabic numerals?

So little is known about this vast domain that our list of open questions could go on and on. With the new brain imaging tools now available, our scientific explorations of the human brain are really just beginning. From neural circuit to mental computation, from single neurons to complex arithmetic functions, cognitive neuroscience has begun to weave increasingly tighter links among brain regions, revealing a more complex and more intriguing picture than we could have imagined. We have only caught the first few glimpses of how neural tissue can become, in the words of Jean-Pierre Changeux and Alain Connes, "matter for thought." Stay tuned, as the next ten years of brain research are most likely to yield many more exciting insights about that special organ that makes us human.

Chapter 9

What Is a Number?

A mathematician is a machine for turning coffee into theorems.

Anonymous

What is a number, that a man may know it, and a man, that he may know a number?" This question, magnificently formulated by Warren McCulloch in 1965, is one of the oldest issues in the philosophy of science—one of those that Plato and his disciples regularly explored on the benches of the first academy twenty-five centuries ago. I often wonder how the great philosophers of the past would have welcomed the recent data from neuroscience and cognitive psychology. What dialogues would the images of positron emission tomography have inspired in Platonists? What drastic revisions would the experiments on neonate arithmetic have imposed on the English empiricist philosophers? How would Diderot have received the neuropsychological data that demonstrates the extreme fragmentation of knowledge in the human brain? What penetrating insights would Descartes have had if he had been fed with the rigorous data of contemporary neuroscience instead of the flights of fancy of his time?

We are close to the end of our exploration into arithmetic and the brain. Now that we have a better grasp of how the human brain represents and manipulates numbers, perhaps we should summarize to what extent these empirical data affect our understanding of the brain and of mathematics. How does the brain acquire mathematics? What is the nature of mathematical intuition, and can one improve it? What are the relations between mathematics and logic? Why is mathematics so efficient in the physical sciences? These are not just the academic ruminations of philosophers hidden in their ivory towers. The answers we give to them have a profound impact on our educational policies and research programs. Piaget's constructivism and Bourbaki's austere rigor have left their marks on our schools. Will such trenchant educational theories ever give way to more serene and better optimized teaching methods, based on a genuine understanding of how the human brain does mathematics? Only a thorough consideration of the neuropsychological bases of mathematics may move us closer to achieving that crucial goal.

Is the Brain a Logical Machine?

What sort of machine is the human brain, that it can give birth to mathematics? Warren McCulloch thought he knew part of the answer. Being a mathematician himself, he was eager to understand "how such a thing as mathematics could have seen the light." As early as 1919, he moved toward the study of psychology and later neurophysiology, with the personal conviction that the brain is a "logical machine." In 1943, in an influential article coauthored with Walter Pitts, he stripped neurons of their complex biological reactions and reduced them to two functions: summing their inputs and comparing this sum to a fixed threshold. He then demonstrated that a network made up of many such interconnected units can perform calculations of an arbitrary complexity. In computer science jargon, such a network has the computational power of a Turing machine—a simple formal device, invented by the brilliant British mathematician Alan Turing in 1937, which captures the essential operations at work in computers for reading, writing, and transforming digital data according to mechanical operations. McCulloch's work thus showed that any operation that can be programmed on a computer can also be performed by an adequately wired network of simplified neurons. In a nutshell, he proclaimed, "A nervous system can compute any computable number."

McCulloch thus followed in the footsteps of George Boole who, in 1854, had set out as a research program for himself "to investigate the fundamental laws of those operations of the mind by which reasoning is performed, to give expression to them in the symbolic language of a calculus, and upon this foundation to establish the science of logic and construct its method."

Boole is the inventor of "Boolean" logic, which describes how the binary values *true* and *false*, denoted by 1 and 0, should be combined in logical computations. Today, Boolean algebra is seen as belonging to mathematical logic or to computer science. But Boole himself considered his research as a central contribution to psychology—an *Investigation of the Laws of Thought,* as his book was titled.

The metaphor of the brain as a computer had now acquired immense popularity, not only with the general public, but even among specialists in cognitive science. It lies at the heart of the so-called "functionalist" approach to psychology, which advocates studying the algorithms of the mind without caring about the workings of the brain. A classical functionalist argument stresses that any digital algorithm computes exactly the same result regardless of whether it runs on a supercomputer or on a pocket electronic calculator. Does it matter, then, that the computer is made of silicon and the brain of nerve cells? For functionalists, the software of the mind is independent of the hardware of the brain—and the mathematical results of Alonzo Church and Alan Turing guarantee that all functions that are computable by a human mind can also be computed by a Turing machine or a computer. In 1983, Philip Johnson-Laird went as far as to state that "the physical nature [of the brain] places no constraints on the pattern of thought," and that as a consequence the brain-computer metaphor "need never be supplanted."

Is the brain really nothing more than a computer or a "logical machine"? Does its logical organization explain our mathematical abilities, and should it be studied independently of its neural substrate? I will not surprise you much if I confess that I suspect that functionalism provides too narrow a perspective on the relations between mind and brain. On purely empirical grounds, the brain-computer metaphor simply does not provide a good model of the available experimental data. The preceding chapters abound in counterexamples that suggest that the human brain does not calculate like a "logical machine." Rigorous calculations do not come easily to *Homo sapiens.* Like so many other animals, humans are born with a fuzzy and approximate concept of number that has little in common with the digital representation of computers. The invention of a numerical language and of exact calculation algorithms belongs to the recent cultural history of humanity—and in several respects, it is an unnatural evolution. Though our culture has invented logic and arithmetic, our brain has remained surprisingly refractory even to the simplest algorithms. By way of proof, one merely needs to consider the difficulty with which children assimilate arithmetic tables and calculation rules. Even a calculating prodigy, after years of training, takes tens of seconds to multiply two six-digit numbers—a thousand to a million times slower than the most sluggish personal computer.

The inadequacy of the brain-computer metaphor is almost comical. In domains in which the computer excels—the faultless execution of a long series of logical steps—our brain turns out to be slow and fallible. Conversely, in domains in which computer science meets its most serious challenges—shape recognition and attribution of meaning—our brain shines by its extraordinary speed.

At the level of the neural circuits themselves, comparing the brain to a "logical machine" does not stand up to scrutiny. Each neuron implements a biological function considerably more complex than the simple logical addition of its inputs (although McCulloch and Pitts's formal neurons sometimes provide a useful approximation to real neurons). Above all, real networks of neurons depart from the rigorous assembly of transistors in the electronic chips of modern computers. Although it is technically possible to assemble formal neurons to build up logical functions, as shown by McCulloch and Pitts, this is not how the central nervous system works. Logical gates are not primitive operations of the brain. If one had to look for a "primitive" function in the nervous system, it would perhaps be the ability of a nerve cell to recognize an elementary "shape" in its inputs by weighing the neuronal discharges it receives from thousands of other units. The recognition of approximate shapes is an elementary and immediate property of the brain, while logic and calculation are derived properties, accessible only to the brain of a single, suitably educated species of primate.

In all fairness, it should be said that many functionalist psychologists do not adhere to the simplistic equation "brain = computer." Their position is more subtle. They do not necessarily identify the brain with any of the serial types of computers that we currently use; but they merely conceive of it as an *information-processing device.* According to them, psychology should be exclusively concerned with the characterization of the transformations that cerebral modules apply to the information they receive. Even if these transformation algorithms are not understood yet, and even if no extant computer is able to implement them, brain functions in principle will eventually be reduced to them. That prospect makes the study of neurons, synapses, molecules, and other properties of the mind's "wetware" irrelevant to psychology.

Even this more subtle brand of functionalism remains questionable, however. Not that it is wrong to study the algorithms of the brain or the activities of humans at a purely behavioral level—one can learn a lot about a machine by determining the fundamental principles on which it is based. But doesn't one make even more progress when one discovers how the machine itself is built? The history of science abounds with examples where the understanding of the physical or biological substrate of a phenomenon has caused a sudden advance in the understanding of its functional properties. The discovery of the molecular structure of DNA, for instance, has radically modified our conception of the "algorithms" of heredity that were discovered years before by Mendel. Likewise, new brain imaging tools are currently revolutionizing our knowledge of cerebral functioning. Wouldn't it be absurd if psychologists were to listen to the functionalists and dismiss these tools as unimportant for our understanding of cognition? As a matter of fact, the vast majority of them, far from from turning their backs on neuroscience research, view it as making a vital contribution to the progress of experimental and clinical psychology.

The functionalists' insistence on the computable aspects of cerebral processing also has another unfortunate consequence. It leads them to neglect other facets of brain function that do not easily fit within the formalism of computer science. This may well be the main reason why cognitive psychology has largely left aside the complex issue of the role of emotions in intellectual life. Yet emotions surely should have a place in any theory of cerebral function, including our present quest for the neural bases of mathematics. Anxiety about mathematics can paralyze children to such an extent that they become unable to acquire even the simplest arithmetical algorithms. Conversely, a passion for numbers can turn a shepherd into a calculating prodigy. In a recent book called *Descartes' Error*, the neuropsychologist Antonio Damasio demonstrates how emotions and reason are tightly linked, to the extent that a lesion of the neural systems responsible for the internal evocation of emotions can have a dramatic impact on the ability to make rational decisions in everyday life. The brain-computer metaphor does not easily put up with such observations, which suggest that cerebral function is not confined to the cold transformation of information according to logical rules. If we are to understand how mathematics can become the object of so much passion or hatred, we have to grant as much attention to the syntax of emotions as to the computations of reason.

Analog Computations in the Brain

The pitfalls of the brain-computer metaphor have not escaped the sagacity of all computer scientists. As early as 1957, John Von Neumann, one of the founding fathers of computer science, said in *The Brain and the Computer,* "The language of the brain [is] not the language of mathematics." Let us not reduce machines to solely digital computers, he recommended. Advanced calculations can be performed by analog machines that ignore mathematical logic entirely. A machine is said to be "analog" when it performs computations by manipulating continuous physical quantities analogous to the variables being represented. In Robinson Crusoe's calculator, for instance, the level of water in the accumulator serves as an analog of number, and addition of water is analogous to numerical addition. Von Neumann had the remarkable insight that the brain is probably a mixed analog-digital machine in which symbolic and analogical codes are seamlessly integrated. Whatever limited abilities our brain exhibits for logic and mathematics may just be the visible result of neural architecture that follows nonlogical rules. In Von Neumann's own words,

> When we talk about mathematics, we may be discussing a *secondary* language, built upon the *primary* language truly used by the central nervous system. Thus, the outward forms of *our* mathematics are not absolutely relevant from the point of view of evaluating what is the mathematical or logical language *truly* used by the central nervous system.

The way in which we compare numbers indeed suggests that we are more similar to an analog machine that to a digital computer. Anyone who writes computer programs knows that the operation of number comparison belongs to the basic set of instructions of the processor. A single calculation cycle of constant duration, often shorter than one microsecond, is enough to assess whether the content of one register is smaller than, equal to, or larger than the content of another. Not so for the brain. In Chapter 3 we saw that an adult takes almost half a second to compare two numbers or any two physical quantities. While a few transistors can implement comparison in an electronic chip, the nervous system has to recruit vast networks of neurons and invest a lot of time to reach the same result.

Moreover, the comparison method that we use is not so easily implemented in a digital computer. Remember that we suffer from a distance effect: It systematically takes us more time to compare two close numbers such as 1 and 2 than two distant numbers such as 1 and 9. In modern computers, by contrast, comparison time is constant regardless of the numbers involved.

Inventing a digital algorithm that reproduces the distance effect is something of a challenge. In a Turing machine, a simple way of coding numbers consists in repeating the same symbol *n* times. Thus, 1 is represented by an arbitrary character *a*, 2 by the string *aa*, and 9 by *aaaaaaaaa*. But the machine can process such strings only character by character. Hence most comparison algorithms respond in a time proportional to the smaller of the two numbers to be compared, totally independent of the distance between them. One can program a Turing machine to count how many symbols distinguish the two numbers, but the simplest algorithm of this kind takes increasingly *less* time as the numbers to be compared get increasingly close, contrary to what holds for the brain.

Binary notation is another simple way of representing numbers in a digital computer. Each number is then coded as a string of bits made up of 0s and *1*s. For instance, 6 is coded as *110*, 7 as *111*, and 8 as *1000*. With an internal code of this kind, however, things take a strange turn: Comparison takes more time for numbers 6 and 7, whose last bit is different, than for the numbers 7 and 8, which differ outright from the first bit. Needless to say, this singular mathematical property finds no echo in psychological observations, which indicate, on the contrary, that 6 and 7 are slightly *easier* to compare than 7 and 8.

Thus, the distance effect, a fundamental characteristic of number processing in the human brain, is not a property that holds of most digital computers. Are there any other types of machines for which a distance effect comes about spontaneously? The answer is yes. Almost any *analog* machine can model the distance effect. Consider the simplest of them: a pair of scales. Place a one-pound weight on the left plate and a nine-pound weight on the right. As soon as you let go, the scales immediately tip to the right, indicating that 9 is larger than 1. Now replace the nine pounds with two pounds and start the experiment again. The scales now hit the right side after a greater length of time. Hence scales, just like brains, find it more difficult to compare 2 and 1 than 9 and 1. Indeed, the time that it takes scales to tip over is inversely proportional to the square root of the difference in weight, a mathematical function that fits nicely with the time it takes us to compare two numbers.

Thus, our mental comparison algorithm can be likened to a pair of scales that "weigh up numbers." The arithmetic abilities of our brain are more easily simulated by an analog machine such as scales than by a digital program. One might object that it is always possible to *simulate* the behavior of an analog device on a digital computer. True enough (although some chaotic physical systems cannot be simulated with absolute precision). But the principles on which the computer is designed then do not capture any significant regularity about the brain: The properties of the system are fully defined by the physical system that one chooses to emulate.

The peculiar way in which we compare numbers thus reveals the original principles used by the brain to represent parameters in the environment, such as a number. Unlike the computer, it does not rely on a digital code, but on a continuous quantitative internal representation. The brain is not a logical machine, but an analog device. Randy Gallistel has expressed this conclusion with remarkable simplicity: "In effect, the nervous system inverts the representational convention whereby numbers are used to represent linear magnitudes. Instead of using number to represent magnitude, the rat [like the *Homo sapiens!*] uses magnitude to represent number."

When Intuition Outruns Axioms

Yet another argument militates against the hypothesis that the brain does mathematics like a "logical machine." Since the end of the nineteenth century, several mathematicians and logicians—Dedekind, Peano, Frege, Russell, and Whitehead among others—have attempted to found arithmetic on a purely formal basis. They designed elaborate logical systems whose axioms and syntactic rules attempted to capture our intuition of what numbers are. However, this formalist approach came up against serious problems that are quite revealing about how difficult it can be to reduce brain function to a formal system.

The simplest of these formalizations of arithmetic was provided by Peano's axioms. Sparing you any mathematical jargon, these axioms essentially reduce to the following statements:

- 1 is a number.
- Every number has a successor, denoted as Sn or simply as $n+1$.
- Every number but 1 has a predecessor (assuming that we consider only the positive integers).
- Two different numbers cannot have the same successor.
- Axiom of recurrence: If a property is verified for number 1, and if the fact that it is verified for n implies that it is also verified for its successor $n+1$, then the property is true of any number n.

These axioms may seem complex and gratuitous. All they do, however, is formalize the very concrete notion of the chain of integers 1, 2, 3, 4, and so on. They satisfy our intuition that this chain has no ending: Any number can always be followed by another number that differs from all the preceding ones. Finally they also allow for a very simple definition of addition and multiplication: Adding a number n means repeating the successor operation n times, and multiplying by n means repeating the addition operation n times.

But this formalism has one major problem. While Peano's axioms provide a good description of the intuitive properties of integers, they also allow for other monstrous objects that we are reluctant to call "numbers," but that satisfy the axioms in every respect. These are called "nonstandard models of arithmetic," and they raise considerable difficulties for the formalist approach.

It is difficult, in only a few lines, to explain what a nonstandard model looks like, but for present purposes a simplified metaphor should suffice. Let us start with the set of usual integers 1, 2, 3, and so on, and let us add other elements that we can picture as being "larger than all other numbers." To the numerical half-line formed by the numbers 1, 2, 3, and so on, let us for instance add a second line spreading toward infinity on both sides:

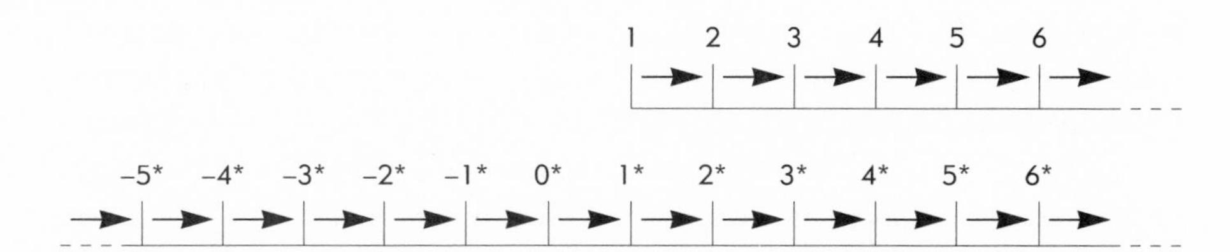

To prevent any confusion, we denote the members of this second number line with a star. Thus –3*, –2*, –1*, 0*, 1*, 2*, 3*, and so on, are all members of this second set. Now let us form the reunion of standard integers and these new elements, and call it the set of "artificial integers":

$$A = \{1, 2, 3, \ldots ; \ldots, -3^*, -2^*, -1^*, 0^*, 1^*, 2^*, 3^*, \ldots\}.$$

Set A truly deserves its name. It is a chimera that does not correspond to anything intuitive. Its elements are the last things that we would want to call "numbers." And yet they verify all of Peano's axioms (with the exception of the axiom of recurrence—this is where my metaphor is oversimplified). Indeed, there is an artificial number 1 that is not the successor of any other artificial number, and every artificial number has a unique and distinct successor in A. The successor of 1 is 2, that of 2 is 3, and so on; and likewise the successor of -2* is -1*, that of -1* is 0*, that of 0* is 1*, and so on. From a purely formal point of view, then, set A provides a fully adequate representation of the set of integers as defined by Peano's axioms—it is a "nonstandard model of arithmetic." In fact, there are an infinity of such models, many of them much more exotic than A.

Nonstandard models are so extravagant that in order to provide a more vivid idea of what they imply, I have to resort to a somewhat farfetched metaphor. In the last century, the classification of animal species seemed well established until a "monster" was discovered in remote Australia: the platypus. Zoologists had not foreseen that some of the criteria they used to classify birds—species having a beak, laying eggs—would also apply to this strange mammal that nobody in the world would want to call a bird. Likewise, Peano could not anticipate that his definition of integers would also apply to mathematical monsters that depart radically from usual numbers.

The discovery of the platypus led zoologists to revise some of their principles. Why wouldn't mathematicians follow their lead? Couldn't they keep adding more axioms to Peano's list until the revised formal system applied to the "true" integers and only to them? We are now reaching the heart of the paradox. A powerful theorem in mathematical logic, first proved by Skolem and deeply related to Gödel's famous theorem, shows that the addition of new axioms can never abolish nonstandard models. As far as they are willing to push the axiomatic formalism, mathematicians will constantly continue to meet new "platypuses"—monsters that will verify all imaginable formal definitions of integers without being identical to them.

In all truth, matters are a trifle more complex because only a certain version of Peano's axioms that mathematicians call "first-order Peano arithmetic" suffers from this infinite expansion of nonstandard models. Yet this version is generally thought to be the best axiomatization of number theory that we have. Thus our best system of axioms fails to capture, in a unique way, our intuitions of what numbers are. The rules behind these axioms seems to fit the "natural" integers tightly; but we later discover that very different objects, which I have called "artificial integers," also verify them. Thus, our "number sense" cannot be reduced to the formal definition provided by these axioms. As was noted by Husserl in his *Philosophy of Arithmetic*, providing a univocal formal definition of what we call numbers is essentially *impossible*: The concept of number is primitive and undefinable.

This conclusion seems implausible. We all have a clear idea of what we mean by an integer, so why should formalizing it be so difficult? Yet all our attempts to provide a formal definition go nowhere. We might try to state, for instance, that integers are obtained by counting: just start with 1 and repeat Peano's "successor" operation as many times as needed. As many times as needed . . . ? But surely not more than a finite number of times; otherwise we would again end up in the strange land of artificial integers! The circularity of the definition becomes obvious: *Numbers* are what one obtains by repeating the successor operation a finite *number* of times.

In *Science and Method*, Poincaré took great pleasure in ridiculing his contemporaries' attempts to define integers through set theory. "Zero is the number of elements in the null class," the mathematician Louis Couturat proposed. "And what is the null class?" Poincaré replied. "It is that class containing no element." Poincaré later charged: "Zero is the number of objects that satisfy a condition that is never satisfied. But as never means *in no case* I do not see that any great progress has been made." Or again, in a biting response to Couturat, who defined 1 as the number of elements of a set in which any two elements are identical: "I am afraid that if we asked Couturat what two is, he would be obliged to use the word one."

Ironically, any five-year-old has an intimate understanding of those very numbers that the brightest logicians struggle to define. No need for a formal definition: We know intuitively what integers are. Among the infinite number of models that satisfy Peano's axioms, we can immediately distinguish genuine integers from other meaningless and artificial fantasies. Hence, our brain does not rely on axioms.

If I insist so strongly on this point, it is because of its important implications for education in mathematics. If educational psychologists had paid enough attention to the primacy of intuition over formal axioms in the human mind, a breakdown without precedent in the history of mathematics might have been avoided. I am referring to the infamous episode of "modern mathematics," which has left scars in the minds of many schoolchildren in France as well as in many other countries. In the 1970s, under the pretext of teaching children greater rigor—an undeniably important goal!—a new mathematical curriculum was designed that imposed a heavy burden of obscure axioms and formalisms on pupils. Behind this educational reform stood a theory of knowledge acquisition that was based on the brain-computer metaphor and that viewed children as little information-processing devices largely devoid of preconceived ideas and capable of ingurgitating any axiomatic system. A group of elite mathematicians known as "Bourbaki" reasoned that teachers should start right away by introducing children to the most fundamental formal bases of mathematics. Indeed, why let pupils lose precious years solving simple, concrete arithmetic problems, when abstract group theory summarizes all such knowledge in a much more concise and rigorous way?

The previous chapters clearly expose the fallacies behind that line of reasoning. The child's brain, far from being a sponge, is a structured organ that acquires facts only insofar as they can be integrated into its previous knowledge. It is well adapted to the representation of continuous quantities and to their mental manipulation in an analogical form. Evolution never prepared it, however, for the task of ingurgitating vast systems of axioms, nor of applying lengthy symbolic algorithms. Thus, quantitative intuition primes over logical axioms. As John Locke astutely observed as early as 1689 in his *Essay on Human Understanding*: "Many a one knows that 1 and 2 are equal to 3 without having thought on any axiom by which it may be proved."

Thus, bombarding the juvenile brain with abstract axioms is probably useless. A more reasonable strategy for teaching mathematics would appear to go through a progressive enrichment of children's intuitions, leaning heavily on their precocious understanding of quantitative manipulations and of counting. One should first arouse their curiosity with some amusing numerical puzzles and problems. Then, little by little, one may introduce them to the power of symbolic mathematical notation and the shortcuts it provides—but at this stage, great care should be taken never to divorce such symbolic knowledge from the child's quantitative intuitions. Eventually, formal axiomatic systems may be introduced.

Even then, they should never be imposed on the child, but rather they should always be justified by a demand for greater simplicity and effectiveness. Ideally, each pupil should mentally, in condensed form, retrace the history of mathematics and its motivations.

Platonists, Formalists, and Intuitionists

We are now ready to discuss McCulloch's second question: "What is a number, that a man may know it?" Twentieth-century mathematicians have been profoundly divided over this fundamental issue concerning the nature of mathematical objects. For some, traditionally labeled "Platonists," mathematical reality exists in an abstract plane, and its objects are as real as those of everyday life. Such was the conviction of Hardy, Ramanujan's discoverer: "I believe that mathematical reality lies outside us, that our function is to discover or *observe* it, and that the theorems which we prove, and which we describe grandiloquently as our "creations," are simply our notes of our observations."

An astonishingly similar profession of faith is found in the French mathematician Charles Hermite: "I believe that the numbers and functions of analysis are not the arbitrary product of our spirits; I believe that they exist outside of us with the same character of necessity as the objects of objective reality; and we find or discover them and study them as do the physicists, chemists, and zoologists."

These two quotations are drawn from Morris Kline's book *Mathematics: The Loss of Certainty*, which contains dozens of similar excerpts. Platonism, indeed, is a prevalent belief system among mathematicians, and I am convinced that it accurately describes their introspection: They really have the *feeling* of moving in an abstract landscape of numbers or figures that exists independently of their own attempts at exploring it. Yet should this feeling be taken at face value, or should we just consider it as a psychological phenomenon that needs to be explained? For an epistemologist, a neurobiologist, or a neuropsychologist, the Platonist position seems hard to defend—as unacceptable, in fact, as Cartesian dualism is as a scientific theory of the brain. Just as the dualist hypothesis faces insurmountable difficulties in explaining how an immaterial soul can interact with a physical body, Platonism leaves in the dark how a mathematician in the flesh could ever explore the abstract realm of mathematical objects. If these objects are real but immaterial, in what extrasensory ways does a mathematician perceive them? This objection seems fatal to the Platonist view of mathematics. Even if mathematicians' introspection convinces them of the tangible reality of the objects they study, this feeling cannot be more than an illusion. Presumably, one can become a mathematical genius only if one has an oustanding capacity for forming vivid mental representations of abstract mathematical concepts—mental images that soon turn into an illusion, eclipsing the human origins of mathematical objects and endowing them with the semblance of an independent existence.

Turning their back on Platonism, a second category of mathematicians, the "formalists," view the issue of the existence of mathematical objects as meaningless and void. For them, mathematics is only a game in which one manipulates symbols according to precise formal rules. Mathematical objects such as numbers have no relation to reality: They are defined merely as a set of symbols that satisfy certain axioms. According to David Hilbert, head of the formalist movement, instead of stating that only one line can go through any two points, one could say that only one table goes through any two glasses of beer—this substitution would not change any of the theorems of geometry! Or according to Wittgenstein's famous statement: "All mathematical propositions mean the same thing, namely nothing."

There is certainly some truth in the formalists' idea that a large part of mathematics is a purely formal game. Indeed, numerous questions in pure mathematics have arisen from what, at first sight, may seem to be fanciful ideas. What would happen if that axiom were replaced by its negation? Or if one turned this "plus" sign into a "minus" sign? Or if taking the square root of a negative number were suddenly allowed? Or if there were integers larger than all others?

And yet I do not believe that the whole of mathematics can thus be reduced to an exploration of the consequences of purely arbitrary choices. Though the formalist position may account for the recent evolution of pure mathematics, it does not provide an adequate explanation of its origins. If mathematics is nothing more than a formal game, how is it that it focuses on specific and universal categories of the human mind such as numbers, sets, and continuous quantities? Why do mathematicians judge the laws of arithmetic to be more fundamental than the rules of chess? Why did Peano go to great pains to propose a few well-chosen axioms rather than a series of haphazard definitions? Why did Hilbert himself select only a restricted subset of elementary numerical reasonings to serve as a tentative foundation for the rest of mathematics? And, above all, why does mathematics apply so tightly to the modeling of the physical world?

I believe that most mathematicians do not just manipulate symbols according to purely arbitrary rules. On the contrary, they try to capture in their theorems certain physical, numerical, geometrical, and logical intuitions. A third category of mathematicians is thus that of the "intuitionists" or "constructivists," who believe that mathematical objects are nothing but constructions of the human mind. In their view, mathematics does not exist in the outside world, but only in the brain of the mathematician who invents it. Neither arithmetic, nor geometry nor logic predate the emergence of the human species. It would even be conceivable for another species to develop radically different mathematics, as Poincaré or Delbrück have suggested. Mathematical objects are fundamental, *a priori* categories of human thought that the mathematician refines and formalizes. The structure of our mind forces us, in particular, to parse the world into discrete objects; this is the origin of our intuitive notions of set and of number.

The founders of intuitionism have stressed the primitive and irreductible nature of numerical intuition. Poincaré spoke about "this intuition of pure number, the only intuition which cannot deceive us," and he confidently proclaimed that "the only natural objects of mathematical thought are the integers." For Dedekind, too, number was an "immediate emanation from the pure laws of thought."

As demonstrated by the mathematics historian Morris Kline, the roots of intuitionism go back to Descartes, Pascal, and of course to Kant. Although Descartes championed the systematic questioning of one's beliefs, he did not go as far as to challenge the obviousness of mathematics. He confessed in his *Meditations*: "I counted as the most certain the truths which I conceived clearly as regards figures, numbers, and other matters which pertain to arithmetic and geometry, and in general to pure and abstract mathematics."

Pascal extended that view even further: "Our knowledge of the first principles, such as space, time, motion, number, is as certain as any knowledge we obtain by reasoning. As a matter of fact, this knowledge provided by our hearts and instinct is necessarily the basis on which our reasoning has to build its conclusions."

For Kant, finally, number belonged to the synthetic *a priori* categories of the mind. More generally, Kant stated that "the ultimate truth of mathematics lies in the possibility that its concepts can be constructed by the human mind."

Among the available theories on the nature of mathematics, intuitionism seems to me to provide the best account of the relations between arithmetic and the human brain. The discoveries of the last few years in the psychology of arithmetic have brought new arguments to support the intuitionist view that neither Kant nor Poincaré could have known. These empirical results tend to confirm Poincaré's postulate that number belongs to the "natural objects of thought," the innate categories according to which we apprehend the world. What, indeed, did the preceding chapters reveal about this natural number sense?

- That the human baby is born with innate mechanisms for individuating objects and for extracting the numerosity of small sets.
- That this "number sense" is also present in animals, and hence that it is independent of language and has a long evolutionary history.

- That in children, numerical estimation, comparison, counting, simple addition and subtraction all emerge spontaneously without much explicit instruction.
- That the inferior parietal region of both cerebral hemispheres hosts neuronal circuits dedicated to the mental manipulation of numerical quantities.

Intuition about numbers is thus anchored deep in our brain. Number appears as one of the fundamental dimensions according to which our nervous system parses the external world. Just as we cannot avoid seeing objects in color (an attribute entirely made up by circuits in our occipital cortex, including area V4) and at definite locations in space (a representation reconstructed by occipitoparietal neuronal projection pathways), in the same way numerical quantities are imposed on us effortlessly through the specialized circuits of our inferior parietal lobe. The structure of our brain defines the categories according to which we apprehend the world through mathematics.

The Construction and Selection of Mathematics

Although the empirical data from neuropsychology seem to provide support for intuitionism, in a form similar to that advocated by Poincaré, this position should be clearly dissociated from an extreme form of intuitionism, the constructivism ardently defended by the Dutch mathematician Luitzen Brouwer. In his zeal to found mathematics on pure intuitions alone, Brouwer went too far, according to many of his colleagues. He took exception to certain logical principles that were very frequently used in mathematical demonstrations but that he felt did not conform to any simple intuition. In particular he was led to reject, for reasons that cannot possibly be explained in full here, the application to infinite sets of the law of excluded middle—an innocent-looking principle of classical logic that states that any meaningful mathematical statement is either true or false. The rejection of that postulate led to the development of a new branch of mathematics called constructivist mathematics.

It is certainly not for me to decide whether classical mathematics or Brouwer's constructivist mathematics provides the most coherent and productive pathways for research. The decision ultimately belongs to the mathematical community, and psychologists must confine themselves to the role of observer. Nevertheless, in my opinion both theories are compatible with the broader hypothesis that mathematics consists in the formalization and progressive refinement of our fundamental intuitions. As humans, we are born with multiple intuitions concerning numbers, sets, continuous quantities, iteration, logic, and the geometry of space. Mathematicians struggle to reformalize these intuitions and turn them into logically coherent systems of axioms, but there is no guarantee that this is at all possible. Indeed, the cerebral modules that underlie our intuitions have been independently shaped by evolution, which was more concerned with their efficiency in the real world than about their global coherence. This may be the reason why mathematicians differ in their choice of which intuitions to use as a foundation and which to relinquish. Classical mathematics are based on an intuition of the dichotomy between truth and falsehood (and as such, as noted by Brouwer, they indeed run the risk of going beyond our intuitions about finite and infinite sets). Brouwer, on the contrary, adopts the primacy of finite constructions or reasonings as a fundamental principle. In the final analysis, his version of mathematics, although it is sometimes called "intuitionism," is certainly not more intuitive than others—it is merely based on a partially distinct set of intuitions.

In this framework, then, what remains to be explained is how, on the basis of the innate categories of their intuition, mathematicans elaborate ever more abstract symbolic constructions. In line with the French neurobiologist Jean-Pierre Changeux, I would like to suggest that an evolutionary process of construction followed by selection is at work in mathematics. The evolution of mathematics is a well-attested fact of history. Mathematics is not a rigid body of knowledge. Its objects and even its modes of reasoning have evolved over the course of many generations. The edifice of mathematics has been erected by trial and error. The highest scaffoldings are sometimes on the verge of collapsing, and reconstruction follows demolition in a never-ending cycle. The foundations of any mathematical construction are grounded on fundamental intuitions such as notions of set, number, space, time, or logic. These are almost never questioned, so deeply do they belong to the irreducible representations concocted by our brain. Mathematics can be characterized as the progressive formalization of these intuitions. Its purpose is to make them more coherent, mutually compatible, and better adapted to our experience of the external world.

Multiple criteria seem to govern the selection of mathematical objects and their transmission to future generations. In pure mathematics, noncontradiction, but also elegance and simplicity, are the central properties that warrant the preservation of a mathematical construction. In applied mathematics, an important criterion is added: the adequacy of mathematical constructs to the physical world. Year after year, mathematical constructions that are self-contradictory, inelegant, or useless are ruthlessly tracked down and eliminated. Only the strongest stand the proof of time.

We first met an example of how selection takes place in mathematics in Chapter 4, when we examined the evolution of number notations. Our remote ancestors probably named only the numbers 1, 2, and 3. Then a series of inventions successively emerged: body-pointing numeration, number names up to ten, and eventually a complex number syntax based on addition and multiplication rules; and in writing, notch-based notation, additive numeration, and eventually positional notation in base 10. Each step saw a small but consistent improvement in the readability, compactness, and expressive powers of numerals.

A similar evolutionary history could be written for the continuum of real numbers. In Pythagoras's time, only integers and ratios of two integers were considered to be numbers. Then came the stupefying discovery of the noncommensurability of the diagonal of the square: $\sqrt{2}$ cannot be expressed as the ratio of two integers. Soon, an infinity of such irrational quantities was constructed. For more than 20 centuries, mathematicians struggled to find a formalism adequate for them. There were false starts—the infinitesimals—apparent solutions that were actually riddled with contradictions, and several returns to square one. Finally, only a century ago, Dedekind's work began to provide a satisfactory definition of the set of real numbers.

According to the evolutionary viewpoint that I defend, mathematics is a human construction and hence a necessarily imperfect and revisable endeavor. This conclusion may seem surprising. Such an aura of purity surrounds mathematics, so often heralded as the "temple of rigor." Mathematicians themselves marvel at the power of their discipline—and rightly so. But don't we all tend to forget that five millennia of efforts have presided over its birth?

Mathematics is often called the only science that is cumulative—its results, once acquired, are never questioned or revised. One look into past mathematics books, however, provides many counterexamples to this view. Monumental volumes have become obsolete with the advent of general methods for solving polynomial equations of the second, third, and fourth degree. A demonstration, which is once found valid, may be judged inadequate or downright false by the next generation of mathematicians. Isn't it amazing, for instance, that the infinite sum $1-1+1-1+1\ldots$, infinitely alternating the addition and subtraction of 1, paralyzed mathematicians for more than a century? Today, any university student can prove that this sum has no meaningful value (it oscillates between 0 and 1). Yet in 1713 a mathematician as talented as Leibniz *proved*—incorrectly, of course—that this infinite sum was equal to ½!

If you find it hard to believe that faulty reasoning can remain hidden from the best minds for decades, take the time to work on the problem depicted in Figure 9.1. It is *proved* in a few steps, that any two lines meet at a right angle! The demon-

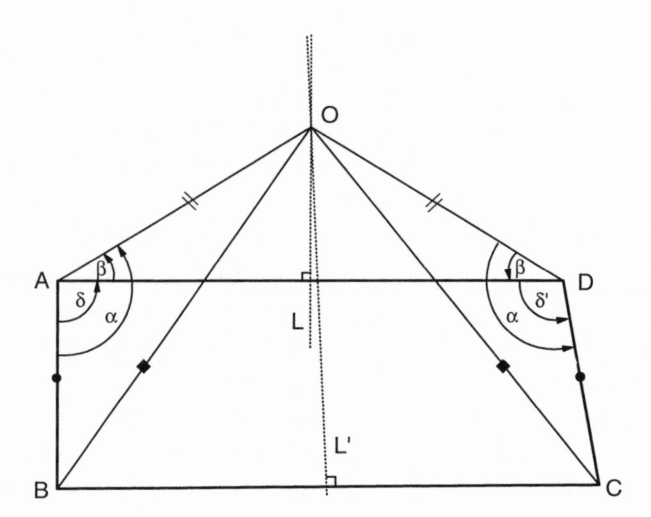

Figure 9.1. The human brain is ill-adapted to the long chains of logical steps required in mathematical demonstrations. In the following proof, although each step seems correct, the final conclusion is obviously wrong since it states that any angle is a right angle! Can you spot the error?

Demonstration: Let ABCD *be a quadrilateral with two equal sides* AB *and* CD *and with a right angle* $\delta = \angle$ BAD. *The angle* $\delta' = \angle$ ADC *is arbitrary—yet we shall prove that it is always equal to the right angle* δ.

Draw L, *the mediator of* AD *and* L', *the mediator of* BC. *Call* O *the intersection of* L *and* L'. *By construction,* O *is equidistant from* A *and* D (OA=OD), *and also from* B *and* C (OB=OC). *Since* AB=CD, *the triangles* OAB *and* ODC *have equal sides and are therefore similar. Hence their angles are equal:* $\angle$ BAO = $\angle$ ODC = α.

Since OAD *is an isosceles triangle,* $\angle$ DAO = $\angle$ ODA = β.

Hence $\delta = \angle$ BAD = $\angle$ BAO − $\angle$ DAO = $\alpha - \beta$; *and* $\delta' = \angle$ ADC = $\angle$ ODC − $\angle$ ODA = $\alpha - \beta$; *which implies that* $\delta = \delta'$. QED.

Where is the error? *See the answer on page 253.*

stration is wrong, of course, but the error is so subtle that it can be sought for several hours without success. What to say, then, of the recent demonstrations that sometimes cover hundreds of pages in mathematical journals? Academies throughout the world have received dozens of false demonstrations of Fermat's last theorem; even the first convincing proof by Andrew Wiles contained an incorrect statement whose rectification took him almost a year of effort. And what are we to think of the newer demonstrations that call for the exhaustive examination of billions of combinations by a computer? Some mathematicians object to this practice, for they fear that we have no proof that the computer program is errorless. To this day, then, the edifice of mathematics is not fully stabilized. We have no guarantee that some of its pieces will not, like Leibniz's infinite sum, be thrown out a few generations from now.

Nobody can deny that mathematics is an extraordinarily difficult activity. I have attributed this difficulty to the architecture of the human brain, which is poorly adapted to long chains of symbolic operations. As children, we already face severe difficulties learning multiplication tables or the multidigit calculation algorithms. Images of cerebral activity during repeated subtractions of digit 3 show intense bilateral activation of parietal and frontal lobes. If an operation as elementary as subtraction already mobilizes our neuronal network to such an extent, one can imagine the concentration and the level of expertise needed to demonstrate a novel and truly difficult mathematical conjecture! It is not so surprising, then, that error and imprecision so often mar mathematical constructions. Only the collective activity of tens of thousands of mathematicians, accumulated and refined over centuries, can explain their present success. This conclusion was aptly captured by the French mathematician Evariste Galois: "[This] science is the work of the human mind, which is destined rather to study than to know, to seek the truth rather than to find it."

The Unreasonable Effectiveness of Mathematics

To affirm that arithmetic is the product of the human mind does not imply that it is arbitrary and that, on some other planet, we might have been born with the idea that 1+1=3. Throughout phylogenetic evolution, as well as during cerebral development in childhood, selection has acted to ensure that the brain constructs internal representations that are adapted to the external world. Arithmetic is such an adaptation. At our scale, the world is mostly made up of separable objects that combine into sets according to the familiar equation 1+1=2. This is why evolution has anchored this rule in our genes. Perhaps our arithmetic would have been radically different if, like cherubs, we had evolved in the heavens where one cloud plus another cloud was still one cloud!

The evolution of mathematics provides some insights into what still stands out as one of mathematics' greatest mysteries: its ability to represent the physical world with a remarkable precision. "How is it possible that mathematics, a product of human thought that is independent of experience, fits so excellently the objects of physical reality?" Einstein asked in 1921. The physicist Eugene Wigner spoke of the "unreasonable effectiveness of mathematics in the natural sciences." Indeed, mathematical concepts and physical observations sometimes seem to fit as tightly as pieces in a jigsaw puzzle. Witness Kepler and Newton discovering that bodies subjected to gravity follow smooth trajectories in the shape of ellipses, parabolas, or hyperbolas—the very curves according to which Greek mathematicians, two millennia earlier, classified the various intersections of a plane and a cone. Witness the equations of quantum mechanics predicting the mass of the electron to the umpteenth decimal. Witness Gauss's bell-shaped curve matching, to near perfection, the observed distribution of the fossil radiation originating from the "Big Bang."

The effectiveness of mathematics raises a fundamental problem for most mathematicians. From their point of view, the abstract world of mathematics should not have to adjust so tightly to the concrete world of physics, because the two are purportedly independent. They perceive the applicability of mathematics as an unfathomable mystery, which leads some of them to mysticism. For Wigner, "the *miracle* of the appropriateness of the language of mathematics to the formulation of the laws of physics is a *wonderful gift* which we neither understand nor deserve." According to Kepler, "the principal object of all research on the external world should be to uncover its order and rational harmony which were *set by God* and which he *revealed* to us in the language of mathematics." Or listen to Cantor: "The highest perfection of God lies in the ability to create an infinite set, and its *immense goodness* leads Him to create it." Ramanujan follows on the same tracks: "An equation for me has no meaning unless it expresses a *thought of God*" (in all these quotations, the emphasis is mine). These statements are not just relics of nineteenth-century mysticism. One version of the anthropic principle, recently adopted by famous contemporary astrophysicists, affirms that the universe was created by design so that humans would eventually emerge from it and be able to understand it.

Was the universe purposely designed according to mathematical laws? It would be foolish to pretend that I can settle an issue that clearly belongs to metaphysics, one that Einstein himself saw as the universe's ultimate mystery. One can at least wonder, however, why eminent scientists feel the need to assert, in the very context of their research, their faith in a universal design and their submission to nonobservable entities, regardless of whether they call them "God" or "the mathematical laws of the universe." In biology, the Darwinian revolution taught us that the finding of organized structures that seem designed for a clear purpose need not point to the works of a Great Architect. The human eye, seemingly a miracle of organization, results from millions of years of blind mutations sorted by natural selection. Darwin's central message is that each time we see evidence for design in an organ such as the eye, we have to ask ourselves whether there ever was a designer or whether selection alone could have shaped it in the course of evolution.

The evolution of mathematics is a fact. Science historians have recorded its slow rise, through trial and error, to greater efficiency. It may not be necessary, then, to postulate that the universe was designed to conform to mathematical laws. Isn't it rather our mathematical laws, and the organizing principles of our brain before them, that were selected according to how closely they fit the structure of the universe? The miracle of the effectiveness of mathematics, dear to Eugene Wigner, could then be accounted for by selective evolution, just like the miracle of the adaptation of the eye to sight. If today's mathematics is efficient, it is perhaps because yesterday's inefficient mathematics has been ruthlessly eliminated and replaced.

Pure mathematics does seem to raise a more serious problem for the evolutionary view I am defending. Mathematicians claim that they pursue some mathematical issues only for beauty's sake, with no applications in sight. Yet decades later their results are sometimes found to fit some hitherto unsuspected problem in physics like a glove. How can one explain the extraordinary adequacy of the purest products of the human mind to physical reality? In an evolutionary framework, perhaps pure mathematics should be compared to a rough diamond, raw material that has not yet been submitted to the test of selection. Mathematicians generate an enormous amount of pure mathematics. Only a small part of it will ever be useful in physics. There is thus an overproduction of mathematical solutions from which physicists select those that seem best adapted to their discipline —a process not unlike the Darwinian model of random mutations followed by selection. Perhaps this argument makes it seem somewhat less miraculous that, among the wide variety of available models, some wind up fitting the physical world tightly.

In the final analysis, the issue of the unreasonable effectiveness of mathematics loses much of its veil of mystery when one keeps in mind that mathematical models rarely agree *exactly* with physical reality. Kepler notwithstanding, planets do not draw ellipses. The earth would perhaps follow an exact elliptic trajectory if it were alone in the solar system, if it was a perfect sphere, if it did not exchange energy with the sun, and so on. In practice, however, all planets follow chaotic trajectories that merely resemble ellipses and are impossible to calculate precisely beyond a limit of several thousand years. All the "laws" of physics that we arrogantly impose on the universe seem condemned to remain partial models, approximate mental representations that we ceaselessly improve. In my opinion, the "theory of everything," the current stuff of physicists' dreams, is unlikely ever to be attained.

The hypothesis of a partial adaptation of mathematical theories to the regularities of the physical world can perhaps provide some grounds for a reconciliation between Platonists and intuitionists. Platonism hits upon an undeniable element of truth when it stresses that physical reality is organized according to structures that predate the human mind. However, I would not say that this organization is mathematical in nature. Rather, it is the human brain that translates it into mathematics. The structure of a salt crystal is such that we cannot fail to perceive it as having six facets. Its structure undeniably existed way before humans began to roam the earth. Yet only human brains seem able to attend selectively to the set of facets, perceive its numerosity as 6, and relate that number to others in a coherent theory of arithmetic. Numbers, like other mathematical objects, are mental constructions whose roots are to be found in the adaptation of the human brain to the regularities of the universe.

There is one instrument on which scientists rely so regularly that they sometimes forget its very existence: their own brain. The brain is not a logical, universal, and optimal machine. While evolution has endowed it with a special sensitivity to certain parameters useful to science, such as number, it has also made it particularly restive and inefficient in logic and in long series of calculations. It has biased it, finally, to project onto physical phenomena an anthropocentric framework that causes all of us to see evidence for design where only evolution and randomness are at work. Is the universe really "written in mathematical language," as Galileo contended? I am inclined to think instead that this is the only language with which we can try to read it.

Appendix

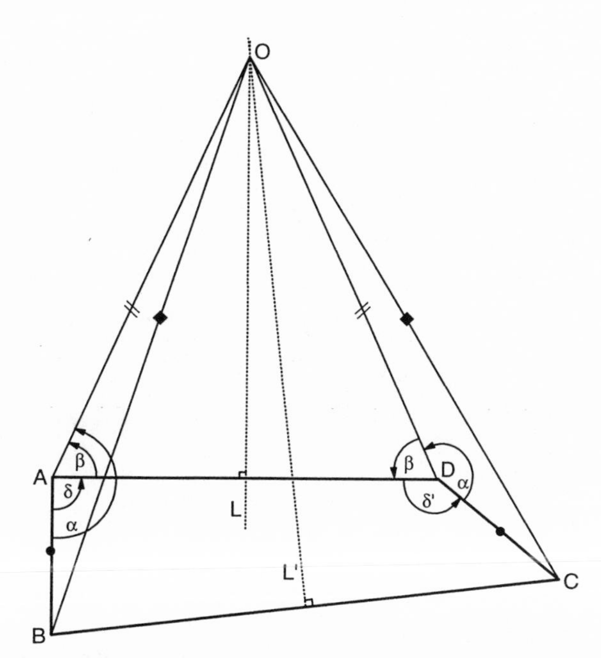

Correction of the "proof" in Figure 9.1. Figure 9.1 was deliberately drawn incorrectly. Though the triangles OAB *and* ODC *are indeed similar, their relations are quite different from those suggested by Figure 9.1 Point* O*, the intersection of* L *and* L'*, is actually much higher (see the above figure). Hence it is true that* $\delta=\alpha-\beta$*, but* $\delta'=2\pi-\alpha-\beta$*. These relations obviously afford no conclusion as to the value of angle* δ'*.*

Notes and References

Chapter 1 Talented and Gifted Animals

The long article by Davis and Pérusse (1988) and the recent books by Gallistel (1990) and Boysen and Capaldi (1993) offer excellent entry points into the literature on animal mathematics. The story of Clever Hans was elegantly narrated by Fernald (1984).

Boysen, S. T., & Berntson, G. G. (1996). Quantity-based interference and symbolic representations in chimpanzees (*Pan trogolodytes*). *Journal of Experimental Psychology: Animal Behavior Processes*, *22*, 76–86.

Boysen, S. T., & Capaldi, E. J. (Eds.) (1993). *The Development of Numerical Competence: Animal and Human Models.* Hillsdale, NJ: Erlbaum.

Capaldi, E. J., & Miller, D. J. (1988). Counting in rats: Its functional significance and the independent cognitive processes that constitute it. *Journal of Experimental Psychology: Animal Behavior Processes*, *14*, 3–17.

Church, R. M., & Meck, W. H. (1984). The numerical attribute of stimuli. In H. L. Roitblat, T. G. Bever & H. S. Terrace (Eds.), *Animal Cognition*. Hillsdale, NJ: Erlbaum.

Davis, H., & Pérusse, R. (1988). Numerical competence in animals: Definitional issues, current evidence and a new research agenda. *Behavioral and Brain Sciences*, *11*, 561–615.

Dehaene, S., & Changeux, J. P. (1993). Development of elementary numerical abilities: A neuronal model. *Journal of Cognitive Neuroscience*, *5*, 390–407.

Fernald, L. D. (1984). *The Hans Legacy: A Story of Science.* Hillsdale, NJ: Erlbaum.

Gallistel, C. R. (1989). Animal cognition: The representation of space, time, and number. *Annual Review of Psychology*, *40*, 155–189.

Gallistel, C. R. (1990). *The organization of learning.* Cambridge, MA: Bradford Books/MIT Press.

Koehler, O. (1951). The ability of birds to count. *Bulletin of Animal Behaviour*, *9*, 41–45.

Matsuzawa, T. (1985). Use of numbers by a chimpanzee. *Nature*, *315*, 57–59.

Mechner, F. (1958). Probability relations within response sequences under ratio reinforcement. *Journal of the Experimental Analysis of Behavior*, *1*, 109–121.

Mechner, F., & Guevrekian, L. (1962). Effects of deprivation upon counting and timing in rats. *Journal of the Experimental Analysis of Behavior*, *5*, 463–466.

Meck, W. H., & Church, R. M. (1983). A mode control model of counting and timing processes. *Journal of Experimental Psychology: Animal Behavior Processes*, *9*, 320–334.

Mitchell, R. W., Yao, P., Sherman, P. T., & O'Regan, M. (1985). Discriminative responding of a dolphin (*Tursiops truncatus*) to differentially rewarded stimuli. *Journal of Comparative Psychology*, *99*, 218–225.

Pepperberg, I. M. (1987). Evidence for conceptual quantitative abilities in the African grey parrot: Labeling of cardinal sets. *Ethology*, *75*, 37–61.

Platt, J. R., & Johnson, D. M. (1971). Localization of position within a homogeneous behavior chain: Effects of error contingencies. *Learning and Motivation*, *2*, 386–414.

Premack, D. (1988). Minds with and without language. In L. Weiskrantz (Ed.), *Thought Without Language*, pp. 46–65. Oxford: Clarenton Press.

Rumbaugh, D. M., Savage-Rumbaugh, S., & Hegel, M. T. (1987). Summation in the chimpanzee (*Pan troglodytes*). *Journal of Experimental Psychology: Animal Behavior Processes*, *13*, 107–115.

Thompson, R. F., Mayers, K. S., Robertson, R. T., & Patterson, C. J. (1970). Number coding in association cortex of the cat. *Science*, *168*, 271–273.

Washburn, D. A., & Rumbaugh, D. M. (1991). Ordinal judgments of numerical symbols by macaques (*Macaca mulatta*). *Psychological Science*, *2*, 190–193.

Woodruff, G., & Premack, D. (1981). Primative (sic) mathematical concepts in the chimpanzee: Proportionality and numerosity. *Nature*, *293*, 568–570.

Chapter 2 Babies Who Count

To the best of my knowledge no exhaustive review has been done of the fast-growing body of literature on numerical abilities in infants. The articles by Wynn (1995), Xu and Carey (1996) and Starkey, Spelke and Gelman (1990) provide excellent starting points.

Antell, S. E., & Keating, D. P. (1983). Perception of numerical invariance in neonates. *Child Development*, *54*, 695–701.

Bijeljac-Babic, R., Bertoncini, J., & Mehler, J. (1991). How do four-day-old infants categorize multisyllabic utterances? *Developmental Psychology*, *29*, 711–721.

Cooper, R. G. (1984). Early number development: Discovering number space with addition and subtraction. In C. Sophian (Ed.), *Origins of Cognitive Skills*, pp. 157–192. Hillsdale, NJ: Erlbaum.

Gelman, R., & Gallistel, C. R. (1978). *The Child's Understanding of Number.* Cambridge, MA: Harvard University Press.

Hauser, M. D., MacNeilage, P., & Ware, M. (1996). Numerical representations in primates. *Proceedings of the National Academy of Sciences USA*, *93*, 1514–1517.

Koechlin, E., Dehaene, S., & Mehler, J. (1997). Numerical transformations in five-month-old human infants. *Mathematical Cognition*, in press.

McGarrigle, J., & Donaldson, M. (1974). Conservation accidents. *Cognition*, *3*, 341–350.

Mehler, J., & Bever, T. G. (1967). Cognitive capacity of very young children. *Science, 158*, 141–142.

Papert, S. (1960). Problèmes épistémologiques et génétiques de la récurence. In Gréco, P., Grize, J.-B., Papert, S., & Piaget, J., *Études d'Épistémologie Génétique. Vol 11. Problèmes de la construction du nombre,* pp. 117–148. Paris: Presses Universitaires de France.

Piaget, J. (1952). *The Child's Conception of Number.* New York: Norton.

Piaget, J. (1954). *The Construction of Reality in the Child.* New York: Basic Books.

Shipley, E. F., & Shepperson, B. (1990). Countable entities: Developmental changes. *Cognition, 34*, 109–136.

Simon, T. J., Hespos, S. J., & Rochat, P. (1995). Do infants understand simple arithmetic? A replication of Wynn (1992). *Cognitive Development, 10*, 253–269.

Starkey, P., & Cooper, R. G., Jr. (1980). Perception of numbers by human infants. *Science, 210*, 1033–1035.

Starkey, P., Spelke, E. S., & Gelman, R. (1983). Detection of intermodal numerical correspondences by human infants. *Science, 222*, 179–181.

Starkey, P., Spelke, E. S., & Gelman, R. (1990). Numerical abstraction by human infants. *Cognition, 36*, 97–127.

Strauss, M. S., & Curtis, L. E. (1981). Infant perception of numerosity. *Child Development, 52*, 1146–1152.

Van Loosbroek, E., & Smitsman, A. W. (1990). Visual perception of numerosity in infancy. *Developmental Psychology, 26*, 916–922.

Wynn, K. (1992). Addition and subtraction by human infants. *Nature, 358*, 749–750.

Wynn, K. (1995). Origins of numerical knowledge. *Mathematical Cognition, 1*, 35–60.

Wynn, K. (1996). Infants' individuation and enumeration of actions. *Psychological Science, 7*, 164–169.

Xu, F., & Carey, S. (1996). Infants' metaphysics: The case of numerical identity. *Cognitive Psychology, 30*, 111–153.

Chapter 3 The Adult Number Line

I have recently reviewed many experiments on elementary number processing in adult humans (Dehaene, 1992, 1993). Seron and collaborators (1992) are largely responsible for renewed interest in "number forms."

Bourdon, B. (1908). Sur le temps nécessaire pour nommer les nombres. *Revue Philosophique de la France et de l'étranger, 65*, 426–431.

Dehaene, S. (1992). Varieties of numerical abilities. *Cognition, 44*, 1–42.

Dehaene, S. (Ed.) (1993). *Numerical Cognition.* Oxford: Blackwell.

Dehaene, S., & Akhavein, R. (1995). Attention, automaticity, and levels of representation in number processing. *Journal of Experimental Psychology: Learning, Memory and Cognition, 21*, 314–326.

Dehaene, S., Bossini, S., & Giraux, P. (1993). The mental representation of parity and numerical magnitude. *Journal of Experimental Psychology: General, 122*, 371–396.

Dehaene, S., & Cohen, L. (1994). Dissociable mechanisms of subitizing and counting—Neuropsychological evidence from simultanagnosic patients. *Journal of Experimental Psychology: Human Perception and Performance, 20*, 958–975.

Dehaene, S., Dupoux, E., & Mehler, J. (1990). Is numerical comparison digital: Analogical and symbolic effects in two-digit number comparison. *Journal of Experimental Psychology: Human Perception and Performance, 16*, 626–641.

Frith, C. D., & Frith, U. (1972). The solitaire illusion: An illusion of numerosity. *Perception & Psychophysics, 11*, 409–410.

Galton, F. (1880). Visualised numerals. *Nature, 21*, 252–256.

Ginsburg, N. (1976). Effect of item arrangement on perceived numerosity: Randomness vs. regularity. *Perceptual and Motor Skills, 43*, 663–668.

Henik, A., & Tzelgov, J. (1982). Is three greater than five: The relation between physical and semantic size in comparison tasks. *Memory and Cognition, 10*, 389–395.

Ifrah, G. (1994). *Histoire universelle des chiffres* (vol. I et II). Paris: Robert Laffont.

Ifrah, G. (1985). *From One to Zero: A Universal History of Numbers.* New York: Viking Press.

Jouette, A. (1996). *Le Secret des Nombres.* Paris: Albin Michel.

Kline, M. (1972). *Mathematical Thought from Ancient to Modern Times.* New York: Oxford University Press.

Kline, M. (1980). *Mathematics: The Loss of Certainty.* New York: Oxford University Press.

Krueger, L. E. (1989). Reconciling Fechner and Stevens: Toward a unified psychophysical law. *The Behavioral and Brain Sciences, 12*, 251–267.

Mandler, G., & Shebo, B. J. (1982). Subitizing: An analysis of its component processes. *Journal of Experimental Psychology: General, 111*, 1–21.

Moyer, R. S., & Landauer, T. K. (1967). Time required for judgements of numerical inequality. *Nature, 215*, 1519–1520.

Ramachandran, V. S., Rogers-Ramachandran, D., & Stewart, M. (1992). Perceptual correlates of massive cortical reorganization. *Science, 258*, 1159–1160.

Seron, X., Pesenti, M., Noël, M. P., Deloche, G., & Cornet, J.-A. (1992). Images of numbers, or "when 98 is upper left and 6 sky blue." *Cognition, 44*, 159–196.

Spalding, J. M. K., & Zangwill, O. L. (1950). Disturbance of number-form in a case of brain injury. *Journal of Neurology, Neurosurgery and Psychiatry, 13*, 24–29.

Trick, L. M., & Pylyshyn, Z. W. (1993). What enumeration studies can show us about spatial attention: Evidence for limited capacity preattentive processing. *Journal of Experimental Psychology: Human Perception and Performance, 19*, 331–351.

Trick, L. M., & Pylyshyn, Z. W. (1994). Why are small and large numbers enumerated differently? A limited capacity preattentive stage in vision. *Psychological Review, 100*, 80–102.

Van Oeffelen, M. P., & Vos, P. G. (1982). A probabilistic model for the discrimination of visual number. *Perception & Psychophysics, 32*, 2, 163–170.

Chapter 4 The Language of Numbers

The history of number notations has been described from slightly different perspectives by Ifrah (1985, 1994), Dantzig (1967) and Hurford (1987). Karen Wynn's 1990 and 1992 articles provide an excellent synthesis of how children acquire number names. The chapter by Ellis (1992) was my main source of references on the impact of number notation on cognitive processing.

Chase, W. G., & Ericsson, K. A. (1981). Skilled memory. In J. R. Anderson (Ed.), *Cognitive Skills and Their Acquisition*, pp. 141–189. Hillsdale, NJ: Erlbaum.

Dantzig, T. (1967). *Number: The Language of Science.* New York, Free Press.

Dixon, R. M. W. (1980). *The Languages of Australia.* Cambridge: Cambridge University Press.

Ellis, N. (1992). Linguistic relativity revisited: The bilingual word-length effect in working memory during counting, remembering numbers, and mental calculation. In R. J. Harris (Ed.), *Cognitive Processing in Bilinguals*, pp. 137–155. Amsterdam: Elsevier.

Fuson, K. C. (1988). *Children's Counting and Concepts of Number.* New York: Springer-Verlag.

Hurford, J. R. (1987). *Language and Number.* Oxford: Basil Blackwell.

Ifrah, G. (1985). *From One to Zero: A Universal History of Numbers.* New York: Viking Press.

Ifrah, G. (1994). *Histoire universelle des chiffres* (vol. 1 and 2). Paris: Robert Laffont.

Marshack, A. (1991). The taï Plaque and calendrical notation in the upper palaeolithic. *Cambridge Archaeological Journal, 1*, 25–61.

Miller, K., Smith, C. M., Zhu, J., & Zhang, H. (1995). Preschool origins of cross-national differences in mathematical competence: The role of number-naming systems. *Psychological Science, 6*, 56–60.

Pollmann, T., & Jansen, C. (1996). The language user as an arithmetician. *Cognition, 59*, 219–237.

Wynn, K. (1990). Children's understanding of counting. *Cognition, 36*, 155–193.

Wynn, K. (1992). Children's acquisition of the number words and the counting system. *Cognitive Psychology, 24*, 220–251.

Zhang, J., & Norman, D. A. (1995). A representational analysis of numeration systems. *Cognition, 57*, 271–295.

Chapter 5 Small Heads for Big Calculations

Gelman and Gallistel's famous book *The Child's Understanding of Number* (1978) remains a central reference on numerical development in children. See also Bideaud (1992), Case (1985, 1992), Fuson (1988) and Hiebert (1986) for more recent references. The research on calculation in adults has been recently summarized by two of its major contributors, Mark Ashcraft (1992, 1995) and Jamie Campbell (1992, 1994). Excellent discussions of issues in mathematical teaching can be found in Baruk (1973, 1985), Paulos (1988), and Stevenson and Stigler (1992).

Ashcraft, M. H. (1992). Cognitive arithmetic: A review of data and theory. *Cognition, 44*, 75–106.

Ashcraft, M. H. (1995). Cognitive psychology and simple arithmetic: A review and summary of new directions. *Mathematical Cognition, 1*, 3–34.

Baroody, A. J., & Ginsburg, H. P. (1986). The relationship between initial meaningful and mechanical knowledge of arithmetic. In J. Hiebert (Ed.), *Conceptual and Procedural Knowledge: The Case of Mathematics*, pp. 75–112. Hillsdale, NJ: Erlbaum.

Baruk, S. (1973). *Echec et maths.* Paris: Editions du Seuil.

Baruk, S. (1985). *L'âge du capitaine.* Paris: Editions du Seuil.

Bideaud, J., Meljac, C., & Fischer, J.-P. (1992). *Pathways to Number.* Hillsdale, NJ: Erlbaum.

Bkouche, R., Charlot, B., & Rouche, N. (1991). *Faire des mathématiques: Le plaisir du sens.* Paris: Armand Colin.

Brown, J. S., & Burton, R. B. (1978). Diagnostic models for procedural bugs in basic mathematical skills. *Cognitive Science, 2*, 155–192.

Campbell, J. I. D. (Ed.) (1992). *The Nature and Origins of Mathematical Skills,* Amsterdam: North Holland.

Campbell, J. I. D. (1994). Architectures for numerical cognition. *Cognition, 53*, 1–44.

Case, R. (1985). *Intellectual Development: Birth to Adulthood.* San Diego: Academic Press.

Case, R. (1992). *The Mind's Staircase: Exploring the Conceptual Underpinnings of Children's Thought and Knowledge.* Hillsdale, NJ: Erlbaum.

Fuson, K. C. (1988). *Children's Counting and Concepts of Number.* New York: Springer-Verlag.

Gelman, R., & Gallistel, C. R. (1978). *The Child's Understanding of Number.* Cambridge, MA: Harvard University Press.

Gelman, R., & Meck, E. (1983). Preschooler's counting: Principles before skill. *Cognition, 13*, 343–359.

Gelman, R., Meck, E., & Merkin, S. (1986). Young children's numerical competence. *Cognitive Development, 1*, 1–29.

Griffin, S., Case, R., & Siegler, R. S. (1994). Rightstart: Providing the central conceptual prerequisites for first formal learning of arithmetic to students at risk for school failure. In K. McGilly (Ed.), *Classroom Lessons: Integrating Cognitive Theory and Classroom Practice*, pp. 25–49. Cambridge, MA: MIT Press.

Groen, G. J., & Parkman, J. M. (1972). A chronometric analysis of simple addition. *Psychological Review, 79*, 329–343.

Hiebert, J. (Ed.) (1986). *Conceptual and Procedural Knowledge: The Case of Mathematics.* Hillsdale, NJ: Erlbaum.

LeFevre, J., Bisanz, J., & Mrkonjic, L. (1988). Cognitive arithmetic: Evidence for obligatory activation of arithmetic facts. *Memory & Cognition, 16*, 45–53.

Lemaire, P., Barrett, S. E., Fayol, M., & Abdi, H. (1994). Automatic activation of addition and multiplication facts in elementary school children. *Journal of Experimental Child Psychology, 57*, 224–258.

Miller, K. F., & Paredes, D. R. (1990). Starting to add worse: Effects of learning to multiply on children's addition. *Cognition, 37*, 213–242.

Paulos, J. A. (1988). *Innumeracy: Mathematical Illiteracy and Its Consequences.* New York: Vintage Books.

Resnick, L. B. (1983). A developmental theory of number understanding. In H. P. Ginsburg (Ed.), *The Development of Mathematical Thinking*, pp. 109–151. New York: Academic Press.

Siegler, R. S., & Jenkins, E. A. (1989). *How Children Discover New Strategies.* Hillsdale, NJ: Erlbaum.

Stevenson, H. W., & Stigler, J. W. (1992). *The Learning Gap.* New York: Simon & Schuster.

Van Lehn, K. (1986). Arithmetic procedures are induced from examples. In *Conceptual and Procedural Knowledge: The Case of Mathematics*, J. Hiebert (Ed.), pp 133–179. Hillsdale NJ: Erlbaum.

Wynn, K. (1990). Children's understanding of counting. *Cognition, 36,* 155–193.

Chapter 6 Geniuses and Prodigies

Binet's book on great calculators, though published in 1894, remains highly readable. One may also profitably consult the more recent books by Smith (1983), Hadamard (1945) and Obler and Fein (1988). Robert Kanigel's 1991 biography of Ramanujan makes for an instructive and fascinating reading. Hundreds of publications on gender differences in mathematics have been reviewed by Benbow (1988) and by Hyde, Fennema, and Lamon (1990).

Benbow, C. P. (1988). Sex differences in mathematical reasoning ability in intellectually talented preadolescents: Their nature, effects, and possible causes. *Behavioral and Brain Sciences, 11,* 169–232.

Binet, A. (1981). *Psychologie des grands calculateurs et joueurs d'échecs.* Paris: Slatkine (original edition: 1894).

Changeux, J. P., & Connes, A. (1995). *Conversations on Mind, Matter, and Mathematics.* Princeton, NJ: Princeton University Press.

Diamond, M. C., & Scheibel, A. B. (1985). Research on the structure of Einstein's brain. In W. Reich, *The Stuff of Genius. The New York Times Magazine*, 28 July 1985, pp. 24–25.

Elbert, T., Pantev, C., Wienbruch, C., Rockstroh, B., & Taub, E. (1995). Increased cortical representation of the fingers of the left hand in string players. *Science, 270,* 305–307.

Flansburg, S. (1993). *Math Magic.* New York: William Morrow & Co.

Geary, D. C. (1994). *Children's Mathematical Development.* Washington DC: American Psychological Association.

Geschwind, N., & Galaburda, A. M. (1985). Cerebral lateralization: Biological mechanisms, associations and pathology. *Archives of Neurology, 42,* 428–259; 521–552; 634–654.

Gould, S. J. (1981). *The Mismeasure of Man.* New York: Penguin.

Hadamard, J. (1945). *An Essay on the Psychology of Invention in the Mathematical Field.* Princeton, NJ: Princeton University Press.

Hardy, G. H. (1940). *A Mathematician's Apology.* Cambridge: Cambridge University Press.

Hermelin, B., & O'Connor, N. (1986). Spatial representations in mathematically and in artistically gifted children. *British Journal of Educational Psychology, 56,* 150–157.

Hermelin, B., & O'Connor, N. (1986). Idiot savant calendrical calculators: Rules and regularities. *Psychological Medicine, 16,* 885–893.

Hermelin, B., & O'Connor, N. (1990). Factors and primes: A specific numerical ability. *Psychological Medicine, 20*, 163–169.

Howe, M. J. A., & Smith, J. (1988). Calendar calculating in "idiots savants": How do they do it? *British Journal of Psychology, 79*, 371–386.

Hyde, J. S, Fennema, E., & Lamon, S. J. (1990). Gender differences in mathematics performance: A meta-analysis. *Psychological Bulletin, 107*, 139–155.

Jensen, A. R. (1990). Speed of information processing in a calculating prodigy. *Intelligence, 14*, 259–274.

Kanigel, R. (1991). *The Man Who Knew Infinity: A Life of the Genius Ramanujan.* New York: Charles Scribner's Sons.

Le Lionnais, F. (1983). *Nombres remarquables.* Paris: Hermann.

Obler, L. K., & Fein, D. (Eds.) (1988). *The Exceptional Brain. Neuropsychology of Talent and Special Abilities.* New York: Guilford Press.

O'Connor, N., & Hermelin, B. (1984). Idiot savant calendrical calculators: Maths or memory? *Psychological Medicine, 14*, 801–806.

Norris, D. (1990). How to build a connectionist idiot (savant). *Cognition, 35*, 277–291.

Sacks, O. (1985). *The Man Who Mistook His Wife for a Hat.* London: Gerald Duckworth & Co.

Schlaug, G., Jäncke, L., Huang, Y., & Steinmetz, H. (1995). In vivo evidence of structural brain asymmetry in musicians. *Science, 267*, 699–701.

Smith, S. B. (1983). *The Great Mental Calculators.* New York: Columbia University Press.

Staszewski, J. J. (1988). Skilled memory and expert mental calculation. In M. Chi, R. Glaser, & M. J. Farr (Eds.), *The Nature of Expertise.* Hillsdale, NJ: Erlbaum.

Thom, R. (1991). *Prédire n'est pas expliquer.* Paris: Flammarion.

Vandenberg, S. G. (1966). Contributions of twin research to psychology. *Psychological Bulletin, 66*, 327–352.

Chapter 7 Losing Number Sense

My colleague Laurent Cohen and I recently published a synthesis of numerous cases of acalculia (Dehaene and Cohen, 1995). See also Deloche and Seron's 1987 book. The recent surge of interest in patients with brain lesions whose number processing is impaired is mostly due to the initial publication of brilliant studies by Alfonso Caramazza and Michael McCloskey; their original papers remain a must in the field.

Anderson, S. W., Damasio, A. R., & Damasio, H. (1990). Troubled letters but not numbers. Domain specific cognitive impairments following focal damage in frontal cortex. *Brain, 113*, 749–766.

Benson, D. F., & Denckla, M. B. (1969). Verbal paraphasia as a source of calculation disturbances. *Archives of Neurology, 21*, 96–102.

Benton, A. L. (1992). Gerstmann's syndrome. *Archives of Neurology, 49*, 445–447.

Caramazza, A., & McCloskey, M. (1987). Dissociations of calculation processes. In G. Deloche & X. Seron (Eds.), *Mathematical Disabilities: A Cognitive Neuropsychological Perspective*, pp. 221–234. Hillsdale, NJ: Erlbaum.

Cipolotti, L., Warrington, E. K., & Butterworth, B. (1995). Selective impairment in manipulating arabic numerals. *Cortex, 31*, 73–86.

Cohen, L., Dehaene, S., & Verstichel, P. (1994). Number words and number non-words: A case of deep dyslexia extending to arabic numerals. *Brain, 117*, 267–279.

Cohen, L., & Dehaene, S. (1995). Number processing in pure alexia: The effect of hemispheric asymmetries and task demands. *NeuroCase, 1*, 121–137.

Cohen, L., & Dehaene, S. (1996). Cerebral networks for number processing: Evidence from a case of posterior callosal lesion. *NeuroCase, 2*, 155–174.

Coslett, H. B., & Monsul, N. (1994). Reading with the right hemisphere: Evidence from transcranial magnetic stimulation. *Brain and Language, 46*, 198–211.

Dehaene, S., & Cohen, L. (1991). Two mental calculation systems: A case study of severe acalculia with preserved approximation. *Neuropsychologia, 29*, 1045–1074.

Dehaene, S., & Cohen, L. (1995). Towards an anatomical and functional model of number processing. *Mathematical Cognition, 1*, 83–120.

Dehaene, S., & Cohen, L. (1997). Cerebral pathways for calculation: Double dissociations between Gerstmann's acalculia and subcortical acalculia. *Cortex*, in press.

Déjerine, J. (1892). Contribution à l'étude anatomo-pathologique et clinique des différentes variétés de cécité verbale. *Mémoires de la Société de Biologie, 4*, 61–90.

Deloche, G., & Seron, X. (Eds.) (1987). *Mathematical Disabilities: A cognitive neuropsychological perspective*. Hillsdale, NJ: Erlbaum.

Gazzaniga, M. S., & Hillyard, S. A. (1971). Language and speech capacity of the right hemisphere. *Neuropsychologia, 9*, 273–280.

Gazzaniga, M. S., & Smylie, C. E. (1984). Dissociation of language and cognition: A psychological profile of two disconnected right hemispheres. *Brain, 107*, 145–153.

Gerstmann, J. (1940). Syndrome of finger agnosia, disorientation for right and left, agraphia and acalculia. *Archives of Neurology and Psychiatry, 44*, 398–408.

Geschwind, N. (1965). Disconnection syndromes in animals and man. *Brain, 88*, 237–294, 585–644.

Grafman, J., Kampen, D., Rosenberg, J., Salazar, A., & Boller, F. (1989). Calculation abilities in a patient with a virtual left hemispherectomy. *Behavioural Neurology, 2*, 183–194.

Greenblatt, S. H. (1973). Alexia without agraphia or hemianopsia. Anatomical analysis of an autopsied case. *Brain, 96*, 307–316.

Hittmair-Delazer, M., Semenza, C., & Denes, G. (1994). Concepts and facts in calculation. *Brain, 117*, 715–728.

Hittmair-Delazer, M., Sailer, U., & Benke, T. (1995). Impaired arithmetic facts but intact conceptual knowledge—a single case study of dyscalculia. *Cortex, 31*, 139–147.

Ingvar, D. H., & Nyman, G. E. (1962). Epilepsia arithmetices: A new physiologic trigger mechanism in a case of epilepsy. *Neurology, 12*, 282–287.

Ionesco, E. *The Lesson*. (translated by Donald M. Allen). English translation copyright © 1958 by Grove Press Inc.

Kopera-Frye, K., Dehaene, S., & Streissguth, A. P. (1996). Impairments of number processing induced by prenatal alcohol exposure. *Neuropsychologia, 34*, 1187–1196.

Luria, A. R. (1966). *The Higher Cortical Functions in Man*. New York: Basic Books.

McCloskey, M., & Caramazza, A. (1987). Cognitive mechanisms in normal and impaired number processing. In G. Deloche & X. Seron (Eds.), *Mathematical Disabilities: A Cognitive Neuropsychological Perspective*, pp. 201–219. Hillsdale, NJ: Erlbaum.

McCloskey, M., Sokol, S. M., & Goodman, R. A. (1986). Cognitive processes in verbal-number production: Inferences from the performance of brain-damaged subjects. *Journal of Experimental Psychology: General, 115*, 307–330.

Senanayake, N. (1989). Epilepsia arithmetices revisited. *Epilepsy Research, 3*, 167–173.

Seymour, S. E., Reuter-Lorenz, P. A., & Gazzaniga, M. S. (1994). The disconnection syndrome: Basic findings reaffirmed. *Brain, 117*, 105–115.

Shallice, T., & Evans, M. E. (1978). The involvement of the frontal lobes in cognitive estimation. *Cortex, 14*, 294–303.

Sokol, S. M, Macaruso, P., & Gollan, T. H. (1994). Developmental dyscalculia and cognitive neuropsychology. *Developmental Neuropsychology, 10*, 413–441.

Temple, C. M. (1991). Procedural dyscalculia and number fact dyscalculia: Double dissociation in developmental dyscalculia. *Cognitive Neuropsychology, 8*, 155–176.

Temple, C. M. (1989). Digit dyslexia: A category-specific disorder in development dyscalculia. *Cognitive Neuropsychology, 6*, 93–116.

Chapter 8 The Computing Brain

Posner and Raichle's 1994 book provides an excellent introduction to brain imaging. A more technical treatment can be found in the volume recently edited by Toga and Mazziotta (1996).

Abdullaev, Y. G., & Melnichuk, K. V. (1996). Counting and arithmetic functions of neurons in the human parietal cortex. *NeuroImage, 3*, 216.

Allison, T., McCarthy, G., Nobre, A., Puce, A., & Belger, A. (1994). Human extrastriate visual cortex and the perception of faces, words, numbers and colors. *Cerebral Cortex, 5*, 544–554.

Changeux, J. P., & Connes, A. (1995). *Conversations on Mind, Matter, and Mathematics.* Princeton, NJ: Princeton University Press.

Dehaene, S. (1995). Electrophysiological evidence for category-specific word processing in the normal human brain. *NeuroReport, 6*, 2153–2157.

Dehaene, S. (1996). The organization of brain activations in number comparison: Event-related potentials and the additive-factors methods. *Journal of Cognitive Neuroscience, 8*, 47–68.

Dehaene, S., Posner, M. I., & Tucker, D. M. (1994). Localization of a neural system for error detection and compensation. *Psychological Science, 5*, 303–305.

Dehaene, S., Tzourio, N., Frak, V., Raynaud, L., Cohen, L., Mehler, J., & Mazoyer, B. (1996). Cerebral activations during number multiplication and comparison: A PET study. *Neuropsychologia, 34*, 1097–1106.

Fuster, J. M. (1989). *The prefrontal cortex* (2nd edition). New York: Raven.

Goldman-Rakic, P. S. (1987). Circuitry of primate prefrontal cortex and regulation of behavior by representational knowledge. In F. Plum & V. Mountcastle (Eds.), *Handbook of Physiology, 5*, 373–417.

Kiefer, M., & Dehaene, S. (1997). The time course of parietal activation in single-digit multiplication: Evidence from event-related potentials. *Mathematical Cognition,* in press.

Lennox, W. G. (1931). The cerebral circulation: XV. The effect of mental work. *Archives of Neurology and Psychiatry, 26*, 725–730.

Posner, M. I., Petersen, S. E., Fox, P. T., & Raichle, M. E. (1988). Localization of cognitive operations in the human brain. *Science, 240*, 1627–1631.

Posner, M. I., & Raichle, M. E. (1994). *Images of Mind.* New York: Scientific American Library.

Roland, P. E., & Friberg, L. (1985). Localization of cortical areas activated by thinking. *Journal of Neurophysiology, 53*, 1219–1243.

Rueckert, L., Lange, N., Partiot, A., Appollonio, I., Litvar, I., Le Bihan, D., & Grafman, J. (1996). Visualizing cortical activation during mental calculation with functional MRI. *NeuroImage, 3*, 97–103.

Sokoloff, L., Mangold, R., Wechsler, R. L., Kennedy, C., & Kety, S. (1955). The effect of mental arithmetic on cerebral circulation and metabolism. *Journal of Clinical Investigations, 34*, 1101–1108.

Toga, A. W., & Mazziotta, J. C. (Ed.) (1996). *Brain Mapping: The Methods.* New York: Academic Press.

Chapter 9 What Is a Number?

Kline (1972, 1980) provides a detailed survey of the history of mathematics. Kitcher (1984) and especially Poincaré (1907a, 1907b) have analyzed with remarkable clarity and coherence the intuitionist and constructivist conceptions of the epistemology of mathematics.

Apéry, R. (1982). Mathématique constructive. In F. Guénard & G. Lelièvre (Eds.), *Penser les mathématiques.* pp. 58–72. Paris: Editions du Seuil.

Changeux, J. P., & Dehaene, S. (1989). Neuronal models of cognitive functions. *Cognition, 33*, 63–109.

Gallistel, C. R. (1990). *The Organization of Learning.* Cambridge, MA: Bradford Books/MIT Press.

Hofstadter, D. R. (1979). *Gödel, Escher, Bach: An Eternal Golden Braid.* New York: Basic Books.

Husserl, E. (1891). *Philosophie der Arithmetik.* Halle: C. E. M. Pfeffer.

Johnson-Laird, P. N. (1983). *Mental Models.* Cambridge, MA: Harvard University Press.

Kitcher, P. (1984). *The Nature of Mathematical Knowledge.* New York: Oxford University Press.

Kline, M. (1972). *Mathematical Thought from Ancient to Modern Times.* New York: Oxford University Press.

Kline, M. (1980). *Mathematics: The Loss of Certainty.* New York: Oxford University Press.

McCulloch, W. S., & Pitts, W. (1943). A logical calculus of the ideas immanent in nervous activity. *Bulletin of Mathematical Biophysics*, *5*: 115–137.

McCulloch, W. S. (1965). *Embodiments of mind.* Cambridge, MA: MIT Press.

Papert, S. (1960). Sur le réductionnisme logique. In Gréco, P., Grize, J.-B., Papert, S. & Piaget, J., *Études d'Épistémologie Génétique. Vol 11. Problèmes de la construction du nombre.* pp. 97–116. Paris: Presses Universitaires de France.

Pélissier, A., & Tête, A. (1995). *Sciences Cognitives: Textes Fondateurs (1943–1950).* Paris: Presses Universitaires de France.

Poincaré, H. (1907). *Science and Hypothesis.* London: Walter Scott Publishing Co.

Poincaré, H. (1907). *The Value of Science.* New York: Science Press.

Von Neumann, J. (1958). *The Computer and the Brain.* New Haven, CT: Yale University Press.

Wigner, E. (1960). The unreasonable effectiveness of mathematics in the natural sciences. *Communications on Pure and Applied Mathematics*, *13*, 1–14.

Index